Science Meets Sports

Science Meets Sports:

When Statistics Are More Than Numbers

Edited by

Christophe Ley and Yves Dominicy

Cambridge
Scholars
Publishing

Science Meets Sports: When Statistics Are More Than Numbers

Edited by Christophe Ley and Yves Dominicy

This book first published 2020

Cambridge Scholars Publishing

Lady Stephenson Library, Newcastle upon Tyne, NE6 2PA, UK

British Library Cataloguing in Publication Data
A catalogue record for this book is available from the British Library

Copyright © 2020 by Christophe Ley, Yves Dominicy and contributors

All rights for this book reserved. No part of this book may be reproduced, stored in a retrieval system, or transmitted, in any form or by any means, electronic, mechanical, photocopying, recording or otherwise, without the prior permission of the copyright owner.

ISBN (10): 1-5275-5856-8
ISBN (13): 978-1-5275-5856-4

We dedicate this book to our families and to the memory of Jacques Dominicy. Christophe Ley especially dedicates this book to his wife Nadine.

Contents

Preface   xiii

1 The Home Run Explosion   1
   1.1 Introduction 2
      1.1.1 Game of Baseball 2
      1.1.2 A Plate Appearance 3
      1.1.3 MLB and the Home Run Committee Report . 4
      1.1.4 Statcast Data 6
      1.1.5 Plan of the Chapter 7
   1.2 Empirical Perspective 7
      1.2.1 Launch Conditions: The RED Zone 7
      1.2.2 Changes in Rates of Batted Balls in the RED Region 8
      1.2.3 Changes in Home Run Rates for Batted Balls in the RED Region 10
   1.3 Modelling Perspective 11
      1.3.1 Introduction 11
      1.3.2 Generalised Additive Model 11
      1.3.3 Estimating Home Run Probabilities 12
      1.3.4 Predicting Home Run Counts 13
   1.4 Conclusions 14

2 Advances in Basketball Statistics   19
   2.1 Introduction 20
   2.2 Basketball Statistics 21
   2.3 Statistical Analyses 23
      2.3.1 Basic Statistical Analyses 24
      2.3.2 Advanced Statistical Analyses 25

	2.4	Statistics Answers Fans' Questions	28
	2.5	Conclusions	43
3	**Measurement**		**53**
	3.1	Introduction—Words and Numbers	54
	3.2	Measurement Best Practices	54
	3.3	Individual Contribution to Team	58
	3.4	Levels of Measurement Redux	59
	3.5	Rating Teams with Disparate Schedules	61
	3.6	Entropy Describes Repertoire	65
	3.7	The Embeddings Approach	66
	3.8	Validity—Words that Matter	72
	3.9	Conclusions	75
4	**Analysing Positional Data**		**81**
	4.1	Introduction	82
	4.2	Related Work	82
	4.3	Player Movement	83
		4.3.1 Positions	83
		4.3.2 Velocities and Accelerations	83
		4.3.3 Movement Models	85
	4.4	Zones of Control	87
	4.5	Intelligent Retrieval	91
	4.6	Conclusions	92
5	**Ranking and Prediction Models**		**95**
	5.1	Introduction	96
	5.2	Modelling of Football Matches	97
		5.2.1 Poisson Models	97
		5.2.2 Ordinal Models	99
	5.3	Ranking Methods	100

		5.3.1	Point-Winning Systems	100
		5.3.2	Least Squares Models	101
		5.3.3	Maximum Likelihood Methods	101
		5.3.4	Elo Models	103
		5.3.5	Examples	104
	5.4	Prediction Models		107
		5.4.1	Poisson Regression	107
		5.4.2	Ordinal Regression	110
		5.4.3	Prediction Measures	111
		5.4.4	Application to the German Bundesliga	112
		5.4.5	Extensions	115
	5.5	Conclusion		117
6	**Running Shoes and Running Injuries**			**123**
	6.1	Introduction		124
	6.2	Part I		125
		6.2.1	It is all in the Study Design	125
		6.2.2	Time to Event Analyses are the Current Gold Standard	128
		6.2.3	Running Shoes do not Cause Running Injuries	130
		6.2.4	Accepting Impermanence: Time-Varying Exposures, Effect Modifiers and Outcomes	131
	6.3	Part II: The Front-Row View		133
		6.3.1	Strong Beliefs and Competition	133
		6.3.2	Running Shoe Prescription does not Work	133
		6.3.3	Is Cushioning Important?	135
		6.3.4	Foot Morphology, Injury Risk and Shoe Types	136
		6.3.5	Other Shoe Features	137
		6.3.6	Throw Your Shoes Away?	137
	6.4	Common Sense Must Prevail		139

Contents

7 Markov Chain Modelling   **147**
- 7.1 Introduction 148
- 7.2 Markov Chain Models 149
  - 7.2.1 Markov Chains 149
  - 7.2.2 The Markov Property 149
  - 7.2.3 Finite Markov Chains 150
- 7.3 Net Games as Finite Markov Chains 153
  - 7.3.1 State-Transition Modelling of Net Games . . . 153
  - 7.3.2 Performance Analysis Using Markov Chains . . 156
  - 7.3.3 Model Validation 157
- 7.4 Game Simulations 159
  - 7.4.1 Simulations Using Changes in Transition Probabilities 159
  - 7.4.2 Simulations Using Numerical Differentiation . . 161
- 7.5 Discussion and Outlook 165
  - 7.5.1 Sports Applications 165
  - 7.5.2 Methodological Outlook 166

8 The Inner Game in Tennis   **171**
- 8.1 Introduction 172
- 8.2 Experimental Studies 173
- 8.3 Qualitative Studies 174
- 8.4 Quantitative Studies 175
  - 8.4.1 Emerging Tools of Quantitative Analysis 177
  - 8.4.2 Identifying Mentality Profiles 178
  - 8.4.3 Emotion Measurement with Computer Vision . 180
- 8.5 Summary 181

9 Tennis Betting Odds   **189**
- 9.1 Introduction 190
- 9.2 Literature Review 191

9.3	Betting Terminology	193
9.4	Testing the Informational Content	198
	9.4.1 Data	198
	9.4.2 Binned Data	198
	9.4.3 Individual Match Data	201
9.5	Practical Implications	206
9.6	Conclusion	208

10 Sports Timetabling — 213

10.1	Introduction	214
10.2	Preliminaries	215
	10.2.1 Sports Timetabling	215
	10.2.2 Fairness Criteria	215
	10.2.3 Fairness Trade-Offs	219
10.3	For Academics	220
	10.3.1 Trade-Offs Between Two Fairness Criteria	220
	10.3.2 Trade-Offs Between Efficiency and Equity	225
10.4	For Practitioners	233
	10.4.1 Trade-Offs Between Two Fairness Criteria	233
	10.4.2 Trade-Offs Between Efficiency and Equity	236
10.5	Conclusion	241

Preface

Aim of the Book

The objective of this book is to present the field of sports statistics to two very distinct target audiences. On the one hand the academicians, mainly statisticians, in order to raise their interest in this growing field, and on the other hand sports fans who, even without advanced mathematical knowledge, will be able to understand the data analysis part and gain new insights into their favourite sports. The book thus offers a unique perspective on this attractive topic by combining sports analytics, data visualisation and advanced statistical procedures to extract new findings from sports data such as improved rankings or prediction methods.

Football, tennis, basketball, track and field, baseball–every sport aficionado should find his/her interest in this book. Thanks to cutting-edge data analysis tools, the present book will provide the reader with completely new insights into his/her favourite sport and this in an engaging and user-friendly way.

Context of the Present Book

The world of sports is currently undergoing a fundamental change thanks to the upcoming trend of sports analytics. Recent advances in data collection techniques have enabled the collection of large, sometimes even massive, amounts of data in all aspects of sports, such as tactics, technique, health complaints and injuries, spatio-temporal whereabouts (e.g., tracking data from GPS), but also marketing and betting. Data are by now regularly collected in almost every sport, ranging from traditional Olympic disciplines to professional football, basketball, and handball, to name but a few. Moreover, massive data from individual recreational athletes, such as runners or cyclists, is available. It is by far not only professional and commercially successful sports clubs that aim to analyse data, even recreational athletes and amateur clubs make use of a variety of sensors to monitor their training and performances.

This global rush towards using advanced statistics and machine learning (or, in modern terms, Data Science) methods in sports is due, in large parts, to the success of the Oakland Athletics baseball team in the 2002 season. Prior to that season, they had hired new players in a till then atypical way, namely by not relying on scouts' experience but rather on sabermetrics, the technical term for empirical/statistical analysis of baseball. This particular story of general manager Billy Beane relying on the use of analytics to assemble a competitive team despite Oakland's small budget has been written up in the famous book *Moneyball: The Art of Winning an Unfair Game* by Michael Lewis in 2003, which was released as movie in 2011 under the title *Moneyball*. The success of the Oakland Athletics team has inspired other teams in baseball and soon after in several other sports. Since then, sports analytics as a field has seen a phenomenal development, having led *inter alia* to the developments of new journals such as the Journal of Sports Analytics whose first edition appeared in 2015.

The present book inscribes itself in this context and aims at further contributing to this stimulating research area thanks to its unique feature of targeting academicians and sports fans.

Content of the Present Book

Various popular sports will be described in this book from a scientific and data-driven perspective. The book covers baseball (Chapter 1 by Albert), basketball (Chapter 2 by Manisera, Sandri and Zuccolotto), both baseball and basketball from the perspective of measurement theory in sports (Chapter 3 by Miller), football (Chapter 4 by Brefeld, Lasek and Mair and Chapter 5 by Groll, Schauberger and Van Eetvelde), running (Chapter 6 by Theisen, Nielsen and Malisoux), net games in general (Chapter 7 by Lames) and tennis in particular (Chapter 8 by Kovalchik and Chapter 9 by Koning and Boot, who focus on betting aspects). Finally, Chapter 10 by Goossens, Yi and Van Bulck covers fairness trade-offs in time-tabling, which is a relevant topic for several sports.

Acknowledgement

We wish to thank all contributors to the present book, which, we hope, will please the reader. We also thank Yvonne Fromme for her professional proof-reading of the entire book. All remaining mistakes are ours.

Chapter 1

The Home Run Explosion

Jim Albert
Bowling Green State University

Abstract

In the game of baseball, many of the runs scored are contributed by home runs. There has been a dramatic increase in the rate of home run hitting in recent seasons, prompting a scientific study to better understand the reasons for the home run increase. Using the new Statcast data, one records the launch velocity and launch angle for every batted ball. Using data from the 2015 through 2019 seasons, this chapter explores the relationship between the launch conditions and home run rates. By use of a generalised additive model, we gain some understanding about the reasons behind the home run increase.

1.1 Introduction

1.1.1 Game of Baseball

Baseball is a bat-and-ball game played by two opposing teams. A game of baseball consists of a series of innings, with each inning consisting of two half-innings. In the top half-inning, the visiting team is batting, and the home team is on the field, and the roles of the two teams are switched for the bottom half-inning.

During a particular half-inning, the pitcher on the fielding team will throw a ball which a player on the opposing team, the batter, will attempt to hit. The intention of the batting team is to have runners advance through four bases (first base, second base, third base, and home plate) to score runs. Each batter will advance to a base or get out, and the half-inning concludes when three outs are recorded. In a professional baseball game, nine innings are played, and the team that scores the most cumulative runs is the winner.

When a new batter comes to bat in a "plate appearance", the pitcher will throw a sequence of pitches. If the batter does not swing at a pitch, then the umpire will call a "strike" or a "ball" depending on the location of the thrown ball. If the batter swings and misses, or if a batter hits the ball in foul territory, a strike is recorded. There are several ways that the plate appearance can conclude. If three strikes are recorded where the last pitch is a called or swinging strike, the batter strikes out. If four balls are recorded or if the batter is hit by the pitch, then the batter can advance to first base. The plate appearance can also end when the batter hits the ball "in-play". When a ball is put in-play, there can be an out (achieved usually by a grounder thrown to first base or a pop-up or fly ball that is caught by a fielder) or a base hit. There are several types of hits (single, double, triple, and home run) distinguished by the number of bases reached by the batter on the hit. The most dramatic hit is the home run, typically achieved when the batter hits a ball a good distance so that it lands over the outfield fence. In this case, the batter, and all runners currently on base will score runs for the batting team. A grand slam home run is a home run hit when all three bases are occupied with runners.

In a typical professional baseball game, a few hundred pitches are thrown, and most of the outcomes of these pitches are strikes or balls, and a relatively small number of balls are put in-play.

1.1.2 A Plate Appearance

Outcomes

A plate appearance (PA) is the basic confrontation between a batter and a pitcher. In a PA there are three basic events that can occur (ignoring other events such as a hit-on-pitch and catcher interference that are unlikely to occur.) The batter can strike out, he can receive a base on balls (called a walk), or he can put the ball in-play. Figure 1 displays the rates of these three events from the 1960 season to the current season (2019). One can see from the figure that there are clear patterns in these rates. The rate of striking out has shown a steady increase in recent years, and the rate of putting a ball in play has steadily decreased. The rate of walking has vacillated over this period of baseball but has shown some increase in recent seasons.

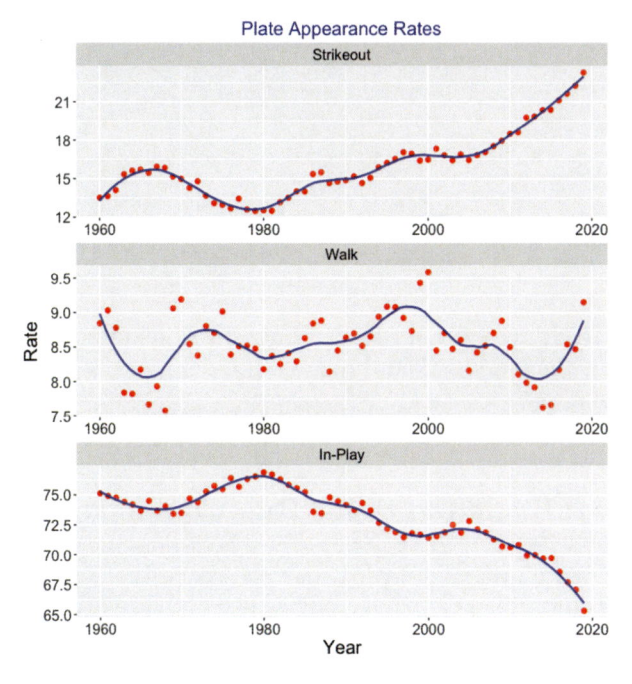

Figure 1.1: Historical pattern of three rates during a plate appearance.

Home Run Rates

The focus of this study is on the rate of home runs per batted ball. Figure 2 displays the home run rate (expressed as a percentage) for the seasons 1960 through the most recent season 2019. From 1960 through 1980 one has seen a decrease in the rate of home run hitting, followed by an increase through 2000, and then a gradual decrease until the 2014 season. There has been a dramatic increase in home run hitting the past five seasons, and the 2019 rate of 5.4% home runs per batted ball is an all-time high.

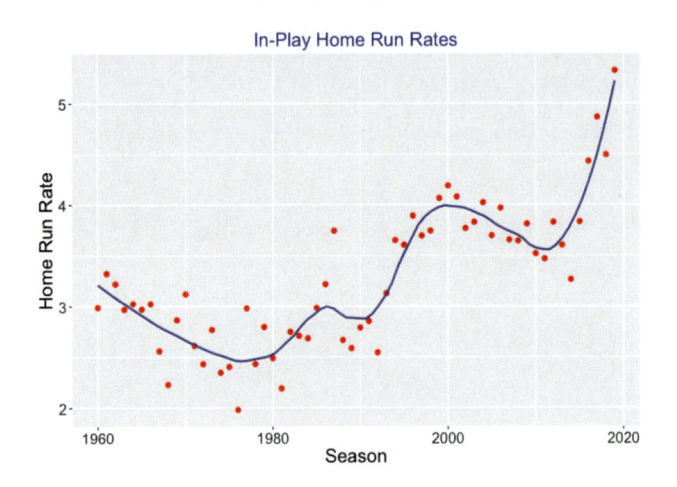

Figure 1.2: Rate of home runs per batted ball (expressed as a percentage) for the seasons 1960 through 2019.

1.1.3 MLB and the Home Run Committee Report

Major League Baseball (MLB) has been concerned about the increase in home run rates. Since home runs are currently prevalent, teams may think of home runs as a primary means of run scoring and fill their batting line-ups with players who are proficient in hitting home runs. An alternative way of scoring runs is based on putting runners on base through base hits or walks and advancing the runners by stolen bases or hits. Due to the home run increase, teams may be less interested in using these "small-ball" methods to score runs. Indeed, one observes, on average, 0.92 stolen bases in the 2019 season which is the smallest average in the past 50 seasons.

1.1. INTRODUCTION

In the fall of 2017, a scientific committee was charged by the Office of the Commissioner of Baseball to "give the full benefit of their knowledge and expertise and to conduct primary and secondary research in order to identify the potential causes of the increase in the rate at which home runs were hit in 2015, 2016, and 2017."

This committee explored several possible explanations for the home run increase.

- **The batters.** It is possible that the batters are hitting balls in a different way that would contribute to the rise in home runs. Perhaps they are hitting balls harder or at a more suitable launch angle or spray angle that would result in more home runs.

- **The pitchers.** Pitchers throw different types of pitches and we have observed a general tendency of the pitch speeds of pitchers to increase in recent years. Perhaps the changes in pitch types and/or pitch speeds are causing the home run increase.

- **The weather.** Baseball is played in a six-month season from April through October and there is a great variation in the weather in the ballpark. It has been documented that it is less likely to hit a home run in cold weather. Perhaps weather changes over recent seasons have contributed to the home run increase.

- **The ballpark.** Every ballpark has a unique shape and ballparks differ in terms of the distance from home plate to the outfield fences. Also, the weather conditions differ among the 30 ballparks. For example, the altitude of Coors Field is 5280 feet and the light air contributes to changes in the movement of a baseball. Perhaps ballpark effects are contributing to the home run changes.

- **The ball.** The composition of the manufactured baseball plays an important role in how the ball moves through the air. It is possible that there have been subtle changes in the ball in recent seasons that have contributed to the home run increase.

The committee explored changes in the launch conditions (the exit velocity, the launch angle, and spray angle) of batted balls over the 2015 to 2017 period. They did not believe that the changes in launch conditions were the primary cause for the increase in home run hitting over this period. Instead, they found that the increase in home runs was primarily due to better "carry" of the baseball for given values of the launch conditions. Furthermore, the committee found that the better carry of the baseball was not due to changes in the weather condition, but instead due to changes in the aerodynamic properties of the baseball. Although the committee believed that changes in the baseball were the main culprit, it was unclear what aspects of the manufactured baseball would lead to a decrease in the drag coefficient and an increase in the ball's carry.

In this chapter we explore this increase in home run hitting given data from the 2015 through 2019 baseball seasons.

1.1.4 Statcast Data

Baseball is remarkable for the amount of data collected on each game. Ever since the beginning of professional baseball in the 19th century, box score data was collected containing the number of at-bats, hits, and runs for each player for every game. (An at-bat is a plate appearance which does not result in a base-on-balls or a hit-by-pitch.) Due to the grassroots efforts of Retrosheet, play-by-play computerised records for every game have been collected, and entire play-by-play records for entire seasons are available at `retrosheet.org`. Starting in 2006, Major League Baseball began to install cameras in every baseball stadium to record information about each pitch. The PitchFX system, created and maintained by Sportsvision, provides information about the speed, movement, and location of every pitch. This data is publicly available and R packages such as pitchRx (Sievert, 2014) allow a person to easily download PitchFX data for a particular group of games.

Statcast (Statcast, 2019) represents the new generation of baseball data. This system was started in 2015 and collects the movements of each player on the baseball field. For a specific player one observes spatial-temporal data–specifically, his location on the field over a fine grid of time values during each inning. In addition, many variables are recorded for each batted ball that is put in-play. One collects the following variables:

- the launch speed, the speed off the bat, measured in miles per hour;

- the launch angle, the angle, measured in degrees, which the ball leaves off from the bat relative to the horizon;

- the spray angle, the horizontal angle of the batted ball relative to home plate.

The complete Statcast dataset is presently only available to the professional teams. However, the batted ball measurements are currently available through the Baseball Savant website at `baseballsavant.mlb.com`. The `baseballr` package of Petti (2019) provides functions for downloading this selected Statcast data over a time period of interest.

For the work of this chapter, Statcast data were collected for all batted balls during the five seasons 2015 through 2019. For each batted ball one collects the launch speed, launch angle, and spray angle. In addition, the data includes the indicator variable `HR` which is equal to one if a home run was observed and equal to zero otherwise.

1.1.5 Plan of the Chapter

The general goal of this chapter is to gain insight about the increase in home run hitting by examining the relationship between the batted ball launch measurements and home run rates. Section 2 takes an exploratory approach where one identifies the launch angle and launch speed measurements that lead to home runs, and one looks at the rate of home run hitting in this region of measurements. We explore how specific rates vary across months of a season and between the 2015 and 2019 seasons. Section 3 uses a modelling approach to see how the probability of a home run depends on launch conditions, month, and season. This modelling approach allows us to predict the home run count in the 2019 season based on the "ball carry" characteristics of previous seasons. In Section 4 we summarise the main findings and discuss related research about the home run increase.

1.2 Empirical Perspective

1.2.1 Launch Conditions: The RED Zone

Following the general strategy in the MLB Home Run Report of Albert et al. (2018), we focus on the values of the launch conditions (launch speed and launch angle) that tend to produce home runs. Figure 3 displays a contour graph of (launch angle, launch speed) values for all home runs hit during the 2019 season and a rectangle is drawn which contains the launch conditions for 78% of the home runs hit for this season. This rectangle is defined by launch angles between 20 and 35 degrees and launch speeds between 98.5 and 108.5 mph. For the remainder of this chapter, we will refer to this region of launch conditions of batted balls as the "RED" region.

We are interested in values of launch conditions among all batted balls that are favourable for hitting home runs. In addition, among all of these "favourable" batted balls, it is of interest to see how many are home runs. This discussion motivates the consideration of the following two rates.

- R_{RED} = fraction of batted balls with launch angle and launch speed measurements in the RED region.

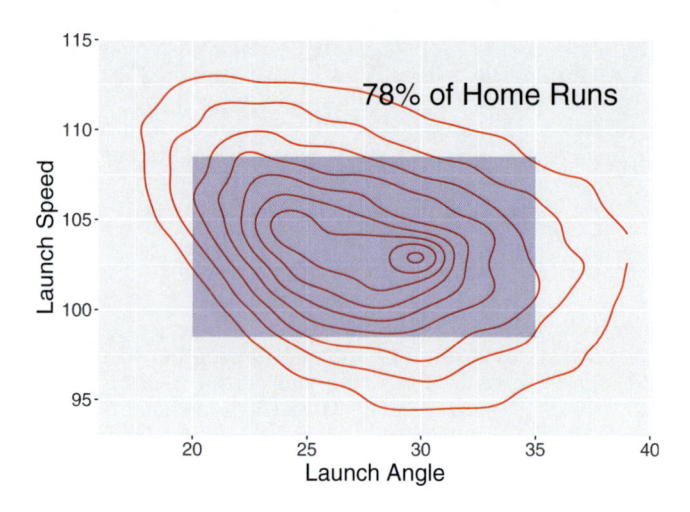

Figure 1.3: Density estimate of the launch angle and launch speed measurements for the home runs hit during the 2019 season. The rectangle contains the values of launch angle and launch speed for which 78 percent of the home runs occur.

- R_{HR} = fraction of balls in this RED region that are home runs.

Both rates are informative about the process of home run hitting. The R_{RED} rate is helpful for understanding possible changes in batting style over seasons. Since players are becoming more familiar with launch angles, it is possible that they will adjust their swing to produce batted balls with suitable launch angles leading to home runs. The R_{HR} rate is helpful for understanding the carry effect of the baseball. If the ball is made in such a way that will change the drag or resistance to the air, this change would result in increased carry and a change in the values of R_{HR}.

1.2.2 Changes in Rates of Batted Balls in the RED Region

Figure 4 graphs the rate of batted balls in the RED region, R_{RED}, for each month of the seasons 2015 through 2019. First, one observes an interesting pattern for the 2015 season–the rate was low in the first three months of the 2015 season but dramatically increased in the second half of the season. Comparing seasons, one observes a

1.2. EMPIRICAL PERSPECTIVE

steady increase in "favourable" batted balls from 2015 through 2019. For example, this RED rate was in the 5 to 5.5 percent range in the 2016-2017 seasons, increasing to 5.5-6.0 in the 2018 season and 6-6.5 in the 2019 season.

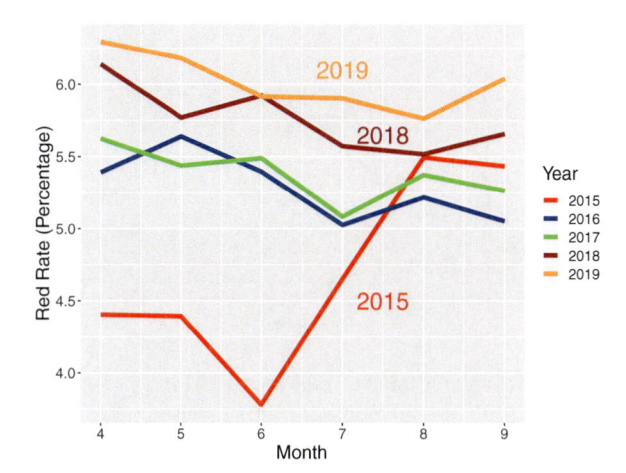

Figure 1.4: Rate of RED zone batted balls for each month of the seasons 2015 through 2019.

Although there is strong evidence for an increase in "home run favourable" batted balls over seasons, there are different populations of hitters for the different seasons. For example, there are rookie players in the 2019 season who did not play in previous seasons and veteran players in earlier seasons who may have retired and did not play in the 2019 season. To control for the changing groups of players, one can focus on particular players who played in both the 2015 and 2019 seasons and see how the favourable batted rates have changed for these players.

We focused on the players who had at least 200 batted balls in 2015 and 100 batted balls in 2019. Figure 5 displays a scatterplot of the RED region rates for these players together with a smoothing curve found using the loess (locally estimated scatterplot smoothing) procedure (Cleveland, 1979). One can compute that 75% of these players had a higher batted ball rate in the RED region in 2019. This indicates that players are indeed changing their swinging style to produce harder hit batted balls with good launch angles. Also, it is interesting that this increase in RED batted ball rates appears to be largest for players with moderate RED rates. The conclusion is that this change in batted ball rates is occurring for players with low, moderate, and high slugging abilities.

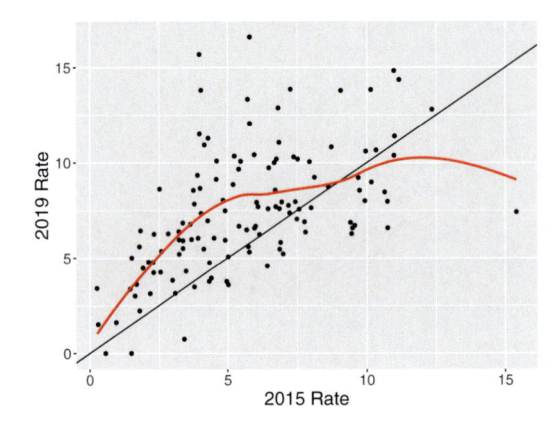

Figure 1.5: Scatterplot of RED region rates for players with large number of batted balls in the 2015 and 2019 seasons.

1.2.3 Changes in Home Run Rates for Batted Balls in the RED Region

Next, we focus on the percentage of batted balls in the RED region that are home runs. Given that a batter has hit a ball with suitable values of launch angle and launch velocity, what is the chance that it will be a home run? This question deals with the characteristic of the baseball to have sufficient carry for a home run.

Figure 6 displays these home run rates over different months and seasons. There is a clear weather effect. Generally, home run rates are smallest in the cold weather month of April and home run rates are larger for the warmer months of June, July, and August.

The pattern of change of home run rates in the RED region over seasons is more complicated. In the 2015 season, the home run rates sharply decreased in the second half of the season. From the second half of the 2015 season through the 2017 season, there was a steady increase in home run rates. This indicates that there was a systematic change in the characteristics of the baseball that led to less drag and an increase in home run rates.

In the last two seasons, we see a different pattern in these home run rates. In 2018, the home run rates in the RED region dramatically decreased and the rates in 2019 resemble the rates in the 2016 season.

1.3. MODELLING PERSPECTIVE

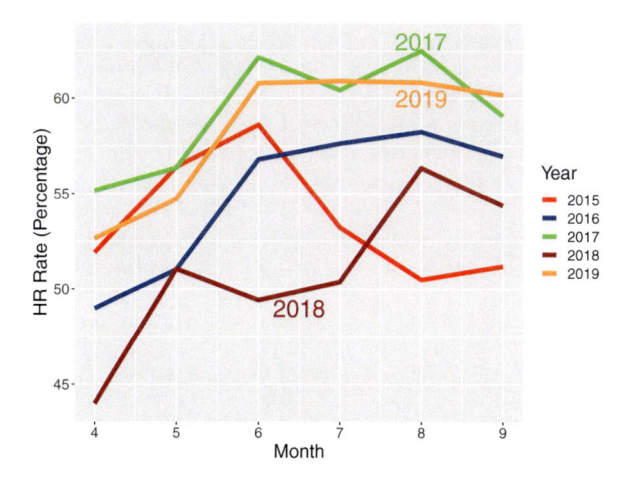

Figure 1.6: Home run rates of batted balls for different months and seasons from 2015 to 2019.

This indicates an increase in drag characteristics of the baseball for 2018 and a decrease in drag in 2019, but not to the level of the 2017 season.

1.3 Modelling Perspective

1.3.1 Introduction

In the empirical approach the RED home run rate was helpful in learning about the likelihood of a home run given suitable values of the launch angle and launch velocity. An alternative approach is to use a statistical model to better understand how the likelihood of a home run depends on the launch conditions. Specifically, one is interested in modelling the probability that a batted ball is a home run based on the launch conditions and season and month effects.

1.3.2 Generalised Additive Model

Let p denote the probability a batted ball is a home run. Suppose we consider the use of the generalised additive model (GAM)

$$\log\left(\frac{p}{1-p}\right) = s(LA, LS) + Season + Month + Season * Month.$$

In this model the term $s(LA, LS)$ denotes a smooth function of the launch angle (LA) and launch speed (LS) and *Season* and *Month* denote categorical effects to the season and month, respectively. It is possible that the month effect depends on the season, so this model includes an interaction effect of season by month.

To demonstrate the usefulness of the nonparametric function $s(LA, LS)$, Figure 7 displays contours of the GAM fitted probability a batted ball is a home run for a region of values of launch angle and launch speed. The contour levels are drawn at fitted probability values of 0.1, 0.3, 0.5, 0.7, and 0.9. In Figure 7, one sees that batted balls hit higher than 100 mph with a launch angle between 25 and 35 degrees tend to be home runs. In addition, note that a higher launch angle compensates for a lower launch speed. For example, the probability of a home run for a batted ball at 30 degrees and 102 mph is approximately equal to the probability of a home run for a batted ball at 20 degrees and 110 mph.

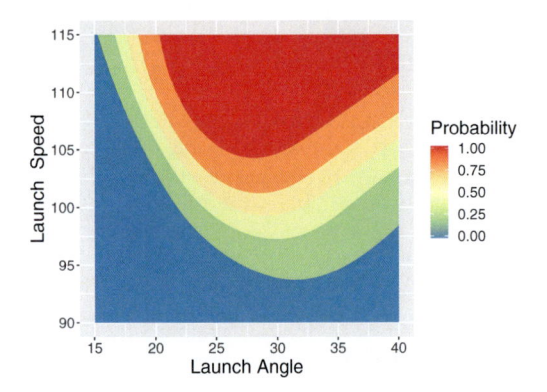

Figure 1.7: Contour graph of probability of home run as a function of the launch angle and launch speed.

1.3.3 Estimating Home Run Probabilities

One use of the GAM model is to estimate the probability a batted ball is a home run at particular launch conditions (launch speed and launch angle) during a particular month and season. This fitted probability is helpful for understanding how the ball carries as a function of launch measurements and specifically how the carry of the ball changes as a function of the month and season.

We focus on particular values of launch angle and launch speed

1.3. MODELLING PERSPECTIVE

that lead to large home run probabilities. Specifically, consider a batted ball hit at a launch angle of 25 degrees and a launch speed of 102 mph. Figure 8 displays the fitted GAM home run probability at these launch conditions for different months and seasons. There are some interesting takeaways from this graph. First, one notices the weather effect–for each season, the smallest home run probability occurs during April, and the probability increases as one moves from April to August. Second, there are substantial differences between the fitted home run probabilities across seasons. Focusing on the home run probability in August (Month = 8), note that there was an increase in the fitted probability from the 2015 to 2017 seasons, but this probability decreased in the 2018 season. These patterns are consistent with the patterns of rates of home runs in the RED region seen in the empirical approach in Figure 6.

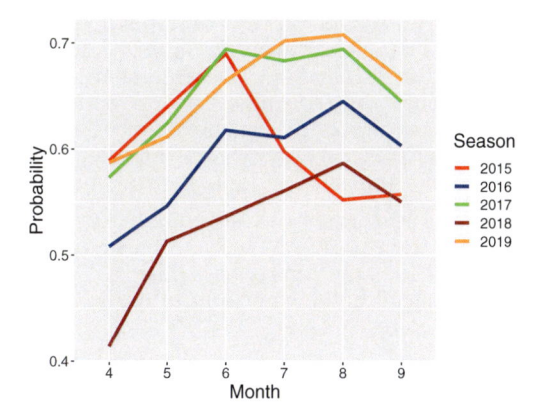

Figure 1.8: Fitted GAM probability of a home run for different months and seasons when the launch angle = 25 degrees and launch speed = 102 mph.

1.3.4 Predicting Home Run Counts

This GAM model can also be used to predict home run counts in future seasons. For example, we observed a surge in home run hitting for the 2019 season. The current home run record was 6105 from the 2017 season and there were 6776 home runs hit in 2019, which broke the season record by 11 percent. Scientists are discussing the reason for this home run surge. Is this surge due to the carry of the ball, or is this surge due to the change in the launch measurements of the hitters?

One can address this question by use of the GAM model. First, we fit this GAM model using all the data from the 2015 through 2018 seasons. Essentially, one is using all the baseballs from the 2015 through 2018 seasons to understand the relationship between launch angle, launch speed, month, and the probability of a home run. Then this fitted GAM model is used to predict the 2019 home run count using the observed 2019 launch condition measurements. If the GAM prediction is close to the actual 2019 home run count, then this tells us that the increase in 2019 home run hitting is due to the changes in launch measurements for the 2019 season. If, instead, our GAM prediction is too small (underestimates the actual 2019 home run count), then that suggests that other inputs, such as the change in the carry of the 2019 balls, are contributing to the 2019 increase.

Using the fitted GAM model, one can obtain a predictive distribution for the 2019 home run rate. Using the launch conditions for each of the batted balls in the 2019 season, one obtains fitted probabilities $\hat{p}$ of a home run for these batted balls. By use of random numbers together with these fitted probabilities, one can predict the total 2019 home run count. By repeating this exercise for 1000 iterations, one obtains a predictive distribution for the home run count.

Figure 9 displays a histogram of the simulated predictions of the 2019 home run count from the GAM model. This figure tells us that the prediction of the 2019 home run count is likely to fall between 6422 and 6593. The actual 2019 home run count of 6776 is represented by a vertical line in the figure.

What does one conclude? One takeaway is that the observed 2019 home run count is larger than the likely range of predicted values. So, this observed home run count is inconsistent with the GAM model based on data from the previous four seasons. This means that one cannot explain the 2019 home run surge solely by the changes in launch conditions in the 2019 hitters. The carry behaviour of the 2015 to 2018 balls (as measured by the GAM fitted model) together with the change in launch measurements in 2019 appear to jointly explain the 2019 home run hitting.

1.4 Conclusions and Problems for Future Study

There has clearly been a dramatic change in home run rates in recent seasons of Major League Baseball, but the reasons for this change are not clear. As described in Section 1, there are many possible explanations for the increase in home runs, such as changes in pitching and batting styles and the composition of the baseball. This chap-

1.4. CONCLUSIONS

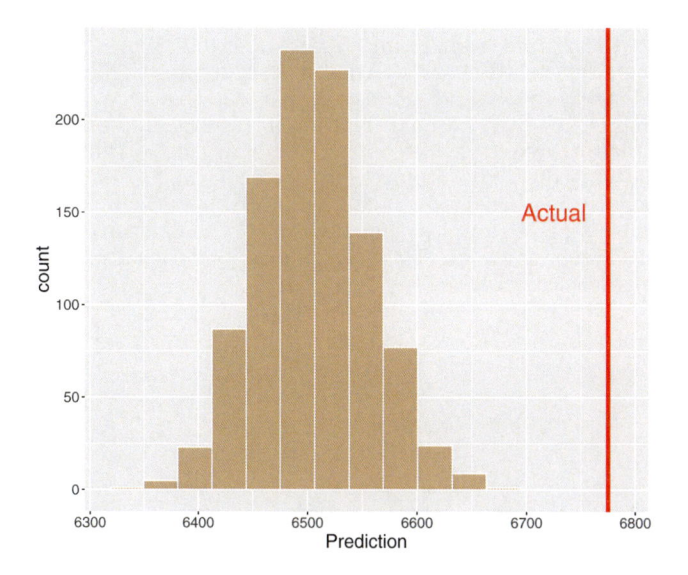

Figure 1.9: Prediction distribution from GAM for home run count in the 2019 season.

ter has focused on the launch conditions of batted balls, specifically the launch angle and launch speed measurements, and the relationship of these launch conditions with home runs. This chapter has demonstrated that the launch conditions of players have changed in recent seasons. Players are generally hitting balls harder and hitting at higher launch angles that would contribute to more home runs. But the composition of the ball appears to play an important role. For example, a contributing factor to the great increase in home run hitting in 2019 compared to 2018 appears to be additional carrying effect of the 2019 baseball.

There is an active effort to learn more about the changes in manufactured baseballs between seasons. For example, Wills (2018) has taken apart baseballs from different seasons and showed how the characteristics of baseballs have changed. Rogers and Ciaccia (2019) describe efforts of Lloyd Smith to measure baseballs with more precision to better understand which characteristics of the baseball would lead to more home runs.

Major League Baseball has been concerned with the increase in home run hitting, thinking that this increase will lead to a fundamental change to baseball and may be less attractive to the fans of the game. In the 2019 season Frank (2019) describes some rule change experiments by the MLB in the Atlantic League (an independent professional league) to see if any changes might lead to a decrease in home run hitting. Currently, baseball plate appearances are dominated by the so-called "three true outcomes" of home runs, strikeouts, and walks that only involve two players–the pitcher and the batter. Many people believe that baseball will be more popular in the future if there are more balls placed in-play that involve all the defensive players in the field. It remains to be seen if the game will eventually move away from the three true outcomes.

Bibliography

Albert, J., Bartroff, J., Blandford, R., Brooks, D., Derenski, J., Goldstein, L., Hosoi, A., Lorden, G., Nathan, A., and Smith, L. (2018). Report of the committee studying home run rates in major league baseball. Technical Report, www.mlb.com.

Cleveland, W. S. (1979). Robust locally weighted regression and smoothing scatterplots. *Journal of the American Statistical Association*, 74:829–836.

Frank, N. (2019). Atlantic league adjusting to, embracing new rules from mlb. https://wtop.com/mlb/2019/07/atlantic-league-adjusting-to-embracing-new-rules-from-mlb.

Petti, B. (2019). Baseballr: Functions for acquiring and analyzing baseball data. R package version 0.5.0. https://billpetti.github.io/baseballr/.

Rogers, J. and Ciaccia, C. (2019). Baseball's home run surge spurs scientists to look for answers. https://www.foxnews.com/science/scientists-search-answers-baseballs-home-run-surge.

Sievert, C. (2014). Taming pitchf/x data with xml2r and pitchrx. *The R Journal*, 6:5–19.

Statcast (2019). https://en.wikipedia.org/wiki/Statcast.

Wills, M. (2018). How one tiny change to the baseball may have led to both the home run surge and the rise in pitcher blisters. https://theathletic.com/381544/2018/06/06/how-one-tiny-change-to-the-baseball-may-have-led-to-both-the-home-run-surge-and-the-rise-in-pitcher-blisters/.

Chapter 2
Advances in Basketball Statistics

MARICA MANISERA
UNIVERSITY OF BRESCIA, ITALY

MARCO SANDRI
UNIVERSITY OF BRESCIA, ITALY

PAOLA ZUCCOLOTTO
UNIVERSITY OF BRESCIA, ITALY

Abstract

In recent years the interest in basketball statistics has greatly increased, and new methods and models have been proposed to analyse basketball data with several interesting aims. In this contribution, after briefly introducing the state of the art of basketball analytics, we offer an overview of possible basic and advanced investigations in performance analysis in basketball, with focus on some opportunities offered by the R package `BasketballAnalyzeR`.

2.1 Introduction

In recent years basketball statistics have become more widespread, and the interest in the applications of statistical methodology to basketball data has increased both among statisticians, who are more focused on the methods, and professionals (coaches, players, scouts, managers, but also fans and sportscasters). A huge number of statistical books and papers have been published on all aspects of basketball, including performance analysis, sports markets, marketing strategies, psychological attributes of players and their impact on match results, medical issues, for example, related to injuries, physical and physiological characteristics of players, fitness and training strategies, scheduling, and many other issues (among others, see Alamar, 2013; Albert et al., 2017; Groll et al., 2018, 2019; Miller, 2015; Passos et al., 2016; Severini, 2014; Winston, 2012; Zuccolotto et al., 2017).

This contribution is focused on performance analysis, which includes both basic tools and more advanced statistical methods. Starting from Dean Oliver's book (Oliver, 2004), the quantitative approach to performance analysis has been applied with a great variety of aims: predicting the result of a match (or a tournament) (Brown and Sokol, 2010; Gupta, 2015; Loeffelholz et al., 2009; Lopez and Matthews, 2015; Manner, 2016; Ruiz and Perez-Cruz, 2015; Vračar et al., 2016; West, 2008; Yuan et al., 2015), investigating the determinants of a team's success (Barber and Rollins, 2019; Csataljay et al., 2009; De Rose, 2004; García et al., 2013; Ibáñez et al., 2009, 2003; Koh et al., 2011, 2012; Sampaio and Janeira, 2003; Trninić et al., 2002), analysing a player's performance and its impact on the team's play (Cooper et al., 2009; Deshpande and Jensen, 2016; Engelmann, 2017; Erčulj and Štrumbelj, 2015; Fearnhead and Taylor, 2011; Franks et al., 2016; Özmen, 2012; Page et al., 2013, 2007; Passos et al., 2016; Piette et al., 2010; Sampaio et al., 2010), analysing the discussed "hot hand" (Arkes, 2010; Avugos et al., 2013; Bar-Eli et al., 2006; Gilovich et al., 1985; Koehler and Conley, 2003; Tversky and Gilovich, 2005; Vergin, 2000), examining performance in high-pressure game situations (Goldman and Rao, 2012; Madden et al., 1995, 1990; Zuccolotto et al., 2018), monitoring playing behaviours, also in order to define new roles (Alagappan, 2012; Bianchi et al., 2017), studying tactics and identifying optimal game strategies (Annis, 2006; McFarlane, 2019; Skinner and Goldman, 2017; Zhang et al., 2013).

With the help of technology, sensor data are now available. This kind of data allows statisticians to extend the range of topics to analyse, for example, the kinetics of body movements, with the aim of identifying the movements leading to the highest shooting efficiency and timing (Aglioti et al., 2008; de Oliveira et al., 2006; Miller and Bartlett, 1996; Okubo and Hubbard, 2006; Zuccolotto et al., 2019). Other interesting topics are the study of players' movements and trajectories and the network of play actions (Ante et al., 2014; Bornn et al., 2017; Bourbousson et al., 2010a,b; Clemente et al., 2015; Fewell et al., 2012; Fujimura and Sugihara, 2005; Gudmundsson and Horton, 2017; Lamas et al., 2011; Metulini et al., 2017a,b, 2018; Miller and Bornn, 2017; Passos et al., 2011; Perše et al., 2009; Piette et al., 2011; Shortridge et al., 2014; Skinner, 2010; Therón and Casares, 2010; Travassos et al., 2012; Wu and Bornn, 2018).

The possibilities in basketball analytics are very wide, and the results that can be obtained from a statistical analysis depend on the available data and software. The aim of this contribution is to give an overview of basic and advanced statistical analyses that can be performed, focusing on the use of the R package BasketballAnalyzeR (Manisera et al., 2019; Sandri, 2020; Zuccolotto and Manisera, 2020). Section 2.2 briefly presents the different data sets existing for statistical analysis in basketball. Section 2.3 offers an overview of possible basic (Subsection 2.3.1) and advanced (Subsection 2.3.2) investigations that can be addressed in basketball analytics, with focus on some of the opportunities offered by BasketballAnalyzeR. Section 2.4 is devoted to answer some of the questions that could be posed by basketball fans.

2.2 Basketball Statistics: State of the Art and Data

The application of statistical methods and models to basketball data yields the so-called basketball analytics. In a schematic way, three main fields exist at the heart of basketball analytics: institutional analyses, sport analytics services, and scientific research.

Institutional analyses are often provided for teams and players in every league or championship, as well as in the lowest level tournaments. They include the number of shots made and attempted, success percentages, number of fouls, rebounds, assists, turnovers, etc. These statistics are generally available for free on the internet and are released in real time or at the end of a game. Basketball fans know them very well, and sport journalists base their articles on these results. However, although useful, institutional analyses are not able to give a relevant contribution in defining winning game strategies or predicting success probabilities.

Sport analytics services are provided by companies specialised in the creation of analysis platforms against payment. These services are often customised based on the customer's needs and give the possibility to record all the events of a match. They return the results of simple statistical analyses, often in the form of graphs and nice visualisation tools that are easy to interpret. Their aim is to offer analytical tools which help the team's staff to make strategic decisions about the next match or identify the best training programmes.

Scientific research deals with basketball analytics with sophisticated analyses, using state-of-the-art statistical techniques and models and developing new methods to address research questions in basketball appropriately. Studies that deeply investigate methodological aspects of basketball statistics cannot provide usable results in real time but require a quite long and complex evaluation, especially in the interpretation of results, even with the help of basketball experts.

In addition, we can also consider analysts working inside basketball teams. Usually, their statistical analyses and results are kept out of the spotlight for fear that they might reveal some secret strategies or weaknesses.

All three fields described above give an important contribution to basketball analytics, addressing several different issues. The range of questions about basketball that may be answered by statistical analyses is growing thanks to the online availability of data sets and increasing computational power. Data can be obtained by multiple sources and according to Zuccolotto and Manisera (2020) can be classified in the following macrocategories: (i) data recorded manually or quasi-manually; (ii) data detected by technological tools; (iii) data from questionnaires; (iv) other data, including, among others, information retrievable from the internet.

Manually (or quasi-manually) data recorded include data from box scores and play-by-play data. Box scores are tables commonly used in basketball that provide a structured summary of the results of a match or a championship. They list the game score as well as individual and team achievements in the game. Typically, they show the number of matches played, the minutes played, the field-goals made and attempted. Those are communicated together with the field goal percentage, three-point and two-point shots attempted and made, with the success percentage, free throws made and attempted and the success percentage, together with some other game variables, like the number of (offensive, defensive and total) rebounds, assists, turnovers, steals, fouls, points scored, etc.

These data are sometimes considered the basketball final statistics to look at. However, from our point of view, they are only the starting point for other analyses. This is a crucial point because we believe that the understanding of basketball cannot be limited to the computation of these numbers. Instead, they must be collected and appropriately structured in order to allow further statistical analyses. Play-by-play data record all the events that occurred during a match; in other words, they provide a transcript of the game as a list of individual events. Each event is described together with the indication of the time of the possession, the player involved in that event, and often including the player's location in that instant.

2.3. STATISTICAL ANALYSES

Examples of data detected by technological tools (category (ii)) include positions of players, referees, and the ball on the court, recorded by sensors or other tracking systems. These kinds of data record defensive and offensive alignment relative to ball location, shot trajectories, player position, and defender proximity to a player.

Data from questionnaires (category (iii)) are usually obtained by means of psychological scales administered to players. The aim is to measure their subjective perceptions and attitudes (for example, leadership, coping strategies, mental toughness, etc.) that can be somehow related to performance.

Category (iv) is a residual category, including, for example, tweets of fans, posts on social media like Facebook or Instagram, and trends of online searches as provided by Google Trends.

While obtaining data of category (iii) and (iv) requires a customised data processing, data belonging to categories (i) and (ii) are often available, with different degrees of quality, in the informative systems of the National and International Federations, sporting organisations, and basketball societies. Basketball data can be available for free or for payment. Usually, when data are free, sophisticated computer tools are needed to obtain the data, such as web scraping procedures.

Data quality is a particularly important issue and the potential of a data set should be assessed with reference to the specific objective of the analysis (Kenett and Redman, 2019). For example, if the focus is on the game evolution in terms of single events, play-by-play data are required. However, a good play-by-play data set must track the game events multiple times at each minute. For example, the play-by-play data currently made available by the Italian Serie A website are aggregated per minute (web.legabasket.it). Without a description second-by-second, no in-depth analyses are viable. In addition, special attention must be paid to the appropriate contextualisation of the obtained results. One must keep in mind that generalisability is not always guaranteed, so, for example, results from NBA data cannot be directly extended to Europe.

2.3 Statistical Analyses Using "BasketballAnalyzeR"

Advances in basketball statistics gain traction from the wide range of questions that arise in the field of basketball analysis. The R package `BasketballAnalyzeR` (Manisera et al., 2019; Sandri, 2020) allows to perform basic and advanced statistical analyses in order to answer several questions. It accompanies the book titled "Basketball Data Science: With Applications in R" (Zuccolotto and Manisera, 2020), which was developed with a substantially empirical approach within the activities of the international network BDsports (Big Data analytics in sports, bdsports.unibs.it). Its main aims include scientific research, education, dissemination, and practical implementation of sports analytics.

2.3.1 Basic Statistical Analyses

In basketball analytics a set of well-known basic indexes and graphs exist that are commonly used by experts, proposed on several specialised websites and well-understood by fans. Examples of **basic indexes** include the Total Basketball Proficiency Score (Kay, 1966), the Individual Efficiency at Games (Gómez Sánchez et al., 1980), and those proposed by Dean Oliver, based on the importance of pace and possessions in the definition of player and team performance and the influence of teamwork on individual statistics. Oliver proposed the definition of offensive and defensive efficiency ratings and the so-called Four Factors (field-goal shooting, offensive rebounds, turnovers, and getting to the free-throw line, Kubatko et al., 2007). They can easily be computed by `BasketballAnalyzeR`.

Several **graphical representations** complete the set of basic analyses. Bar-line plots, radial plots, scatter plots, and bubble plots can give interesting insights on the characteristics of games, teams, and players. For example, a bar-line plot helps to compare the offensive statistics of several teams or players; radial plots can be useful to represent a profile, based on several game features, for each team or player considered. Scatterplots suggest relationships between two variables (or even three if points are colour- or symbol-coded), measured on the teams or players, and allow to identify possible anomalous situations. Bubble plots even represent four characteristics of teams or players in one single plot since size and colour of bubbles vary according to two variables that are added to the two that define x and y location. For example, a bubble plot of a team's players can be defined using 2-point and 3-point shot percentages on the x and y axes, respectively, and bubbles can be coloured according to the scoring percentage of free throws and dimensioned according to the total number of attempted shots. Shot charts are widely employed in basketball analytics. They give clear indications on the players' positions on the court during the match and can be enriched with useful statistics that help understanding what happens on the court.

Other simple tools of descriptive statistics can be used to analyse **performance variability**. Variability indexes and plots inform us about the extent to which players or teams perform differently from each other. Special caution should be used in interpreting variability results. For example, if we are interested in evaluating performance variability of the players in one team, high variability indicates the presence of some great players and some bad ones with reference to

2.3. STATISTICAL ANALYSES

the game variable analysed. If the analysed game variable relates to a specific task (for example, number of assists), high variability suggests that some of the players specialise in assisting their teammates and the team balance is good. On the contrary, if the analysed variable refers to a generic task (for example, field goal percentages), a high variability is a sign of high dependency of the team on a few players whose performance is higher than average.

BasketballAnalyzeR also allows us to run an **inequality analysis** within a basketball team. Borrowed from the economics field, inequality analysis in basketball evaluates the extent to which the distribution of some performance measure (for example, the number of points made by one team) deviates from a perfectly equal distribution (all the players score the same number of points) and from a maximally unequal distribution (one single player scores the total number of points). The value of the Gini's inequality coefficient[1] and the graphical representation of the Lorenz curve [2] allow comparisons among the inequality of several teams.

2.3.2 Advanced Statistical Analyses

Numerous advanced statistical analyses exist that can be performed in basketball studies with a variety of aims. The long, although non-exhaustive, list of papers cited in Section 2.1 gives an idea of the recent publications on this topic. Some of them are directly available in BasketballAnalyzeR while others can use the results from the package as a starting point to develop further methodologies and applications.

A first set of analyses can be carried out to study the **association among variables**. In a broad sense, one can investigate statistical de-

---

[1] The Gini's inequality index is a normalised index measuring the degree of inequality in a given distribution. It ranges from 0 (perfect equality) to 1 or 100% (maximal inequality). In the case of income distribution of a nation's N residents, a value of 0 indicates that all residents have the same income while a value of 1 is obtained when all of the nation's income is owned by only one person.

[2] The Lorenz curve graphically represents the fraction y of the total variable (for example, income) that is cumulatively referred to the bottom fraction x of the population. The two extreme situations of perfect equality and maximal inequality are represented by the straight line $y = x$ and the line having $y = 0$ for all $x \leq (N-1)/N$ and $y = 1$ when $x = 1$, respectively. The observed Lorenz curve lies in between these two extremes. The closer the curve is to the perfect equality line, the smaller the inequality level. The Gini coefficient is computed as the ratio of the inequality area (that is the area between the perfect equality line and the observed Lorenz curve) to the maximum inequality area (given by the area between the perfect equality and the perfect inequality line).

pendence (for example, to measure how much the number of rebounds by one team depends on the opposing team), mean dependence (to evaluate, for example, if the average number of points scored by all the NBA teams differs between the East and West conferences), and correlation. In particular, pairwise linear correlation among variables can be studied creating a correlation matrix and its graphical representation in order to examine the degree, direction, and significance of linear relationships between game variables measured on the single players.

The **similarity among teams or players** with respect to selected game variables can also be assessed and graphically represented by a specific function of BasketballAnalyzeR by resorting to the multivariate data analysis technique of Multidimensional Scaling. It reduces a high-dimensional data set (several teams or players on which a high number of variables have been measured) into a low-dimensional map (usually two dimensions are retained), displaying teams or players as points: points close to each other have similar characteristics, while distant points indicate peculiar teams or players. The dimensionality reduction implies a loss of information which must be evaluated in order to assess the quality of the resulting representation (the stress index is usually employed).

An important issue in basketball statistics is the study of the relationships among players and the impact of their interaction on the team's achievements. Several statistical methods can be used to this aim, for example, **network analysis**, in the sense of the statistical analysis of network data. In basketball analytics, the system conceptualised as the network is the team, and the focus is on modelling the complex statistical dependencies among players, often using high-dimensional data, with the final aim of predicting the team behaviour. For example, BasketballAnalyzeR allows to investigate the network of assists in a team using play-by-play data. The resulting graph is a net displaying the players who mostly interact as those who make and receive most assists. This issue can then be further investigated by considering the play in the absence of some key players in order to examine how the team reorganises its game strategy.

The frequency of occurrence of some events with respect to some variable of interest can give interesting insights on the way a certain team or player is playing. For instance, it can be interesting to examine the **frequency of shots in time** (e.g., with respect to the seconds played in a quarter) **or in space** (e.g., looking at the shot distance) in order to investigate the players' performance in specific moments or areas. Statistics give a response to these questions by density estimation methods, including several tools like histograms, naive, kernel, nearest neighbour, maximised penalised likelihood, etc. (Silverman, 1986). It becomes clearer that, for instance, if the considered players tend to concentrate their shots for a particular moment of the match or in a given area of the court, then their scoring probability varies across time and space, giving useful information to define winning game strategies and improve training programmes.

Among the data mining algorithms allowing advanced statistical analyses in basketball we can mention **cluster analysis**, a broad methodology including several techniques which differ remarkably in their functioning. Cluster analysis is an unsupervised classification method that aims at grouping players, teams, matches, game moments, etc. into classes not a priori defined. Clusters gather observation units that are similar to each other while different from units belonging to the other clusters. The analysis of the goodness of the solution together with the clusters' profiles and characteristics can give interesting insights on the structure of the data set and finally on the phenomenon under study. In basketball, clusters of players can be used to identify similar players and re-define their roles in playing that can differ from the traditional positions (among others, Bianchi et al., 2017). Indeed, the historical five positions (point guard, shooting guard, centre, small forward, and power forward) were defined a long time ago when even the basketball rules were different (for example, the introduction of the 3-point line dates back to the early 80's). New rules have led to changes in the players' physical preparation, their playing style, and the way players interpret their role. In the end, new positions can reflect updated points of view about the game and can be identified by means of statistical analyses based on game variables. Also, matches can be clustered based on their ease. Afterwards, further analyses can be carried out on the obtained clusters with the final aim of identifying losing and winning factors (Lorenzo et al., 2010). Game moments can also be classified, for example, according to the game schemes adopted by players, in order to analyse the play style of a team during a match (Metulini et al., 2018).

Naturally, advances in basketball statistics also concern the very broad field of **statistical models** in their wide meaning of both models with mathematical formalisation and algorithmic models. Every objective in basketball statistics can be achieved by using appropriate models, from the simplest ones as linear regressions and nonparametric regressions, to the most complicated ones being able to model multivariate nonlinear relationships, time dependencies, interactions among variables, and observation units, different nature of the involved variables, etc. A non-exhaustive literature review can be found, for example, in Zuccolotto and Manisera (2020), where also some recent scientific papers are discussed in detail, namely a study on the scoring probability in the presence of high-pressure game situations (Zuccolotto et al., 2018), the definition of new roles in basketball (Bianchi et al., 2017), the analysis of players' movements, and their effect on the team performance (Metulini et al., 2018).

2.4 Statistics Answers Fans' Questions

This section is devoted to answer some of the possible basketball fans' questions, with the aim of showing how basketball statistics "are more than numbers". The R package BasketballAnalyzeR (Manisera et al., 2019; Sandri, 2020) is really helpful in fostering the meeting between statistics and basketball because it can be fruitfully used by interested users without a strong statistical background.

This Q&A section, with examples of questions from fans with the corresponding answers by basketball analysts, is developed using the data from the regular season (82 games) of the NBA championships 2018/2019 and 2017/2018. In much detail, we used one data set including additional information (for example, conference, division, etc.) and three box scores data (2018/2019): the teams' box scores, containing the achievements of all the analysed teams, the opponents' box scores for the achievements of the opponents of each of the analysed teams, and the players' box scores for the individual achievements of each single player in the considered games. In addition, we have play-by-play data recording the events of the 82 games played by the Cleveland Cavaliers during the NBA regular season 2017/2018.

Q1: What were offensive and defensive performance of the four NBA Conference finalists?

Although a universal definition of performance does not exist, we can refer to the famous Four Factors by Dean Oliver (Kubatko et al., 2007) as a good starting point to measure offensive and defensive performance of one team or one player. The Four Factors are (1) the Effective Field Goal Percentage (eFG%); (2) the Turnovers per possession (TO Ratio); (3) the Rebounding Percentages (REB%) and (4) the Free Throw Rate (FT Rate). They can be computed for both offence and defence thus giving a measure of offensive and defensive performance from four points of view. Applying the function fourfactors of BasketballAnalyzeR to the data of the four 2018/2019 Conference finalists (Milwaukee Bucks, Toronto Raptors, Golden State Warriors, and Portland Trail Blazers), Figure 2.1 is obtained. It shows the four factors for every team besides information about pace, possessions, and offensive and defensive ratings.

In detail, Figure 1 shows that the pace of the games increases moving from Portland Trail Blazers to Toronto Raptors, Golden State Warriors and Milwaukee Bucks. The Golden State Warriors have the best offensive performance (the highest offensive rating), while the Milwaukee Bucks are the best team in defence (they have the lowest defensive rating, that is, the offensive ratings of the opponents). The bars in the bottom plots represent the Four Factors: each bar

2.4. STATISTICS ANSWERS FANS' QUESTIONS

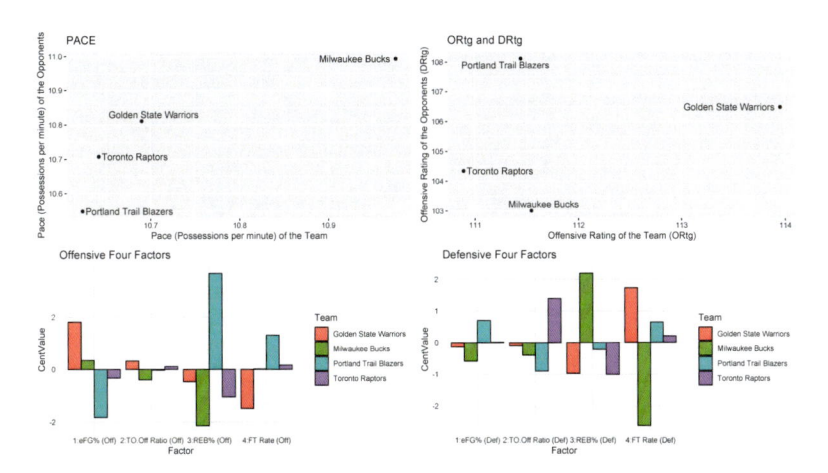

Figure 2.1: Pace, Offensive/Defensive Ratings and Four Factors (differences between the team and the average of the considered teams) - Conference finalists 2018/2019.

measures, for each team and each Factor, the difference between the team value and the average of the four analysed teams. A positive (negative) bar indicates a value above (below) the mean and suggests a strength or a weakness of the team (depending on which of the Four Factors is referred to, as explained later). The Portland Trail Blazers show a particularly good performance in offensive rebounds and free-throw rate, with a performance lower than the average on the effective field goal percentage. This picture is completed, on the defensive side, by a bad performance in the effective field goal percentage and free throw rate. This result is consistent with the fact that the Portland Trail Blazers lost the Western Conference final versus the Golden State Warriors (obviously, there are many other determinants of the outcome of a match).

Q2: Which NBA Eastern conference team had the best performance according to defensive statistics?

It is possible to plot the main defensive statistics of the Eastern Conference teams in one single graph. With the function `barline` we obtain Figure 2.2, where steals, blocks and defensive rebounds are represented on the bars, which are ordered (in decreasing order) according to the points scored by the opponents; the grey line measures the opponents' turnovers (TOV.O), whose scale is on the right vertical axis.

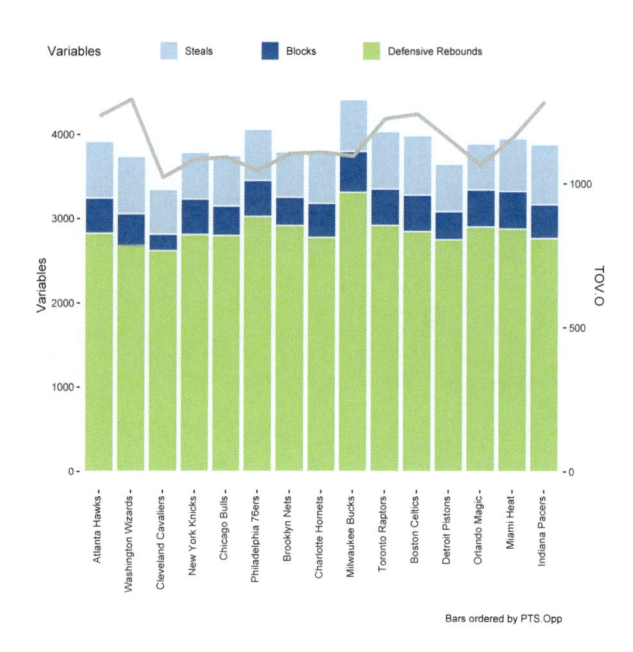

Figure 2.2: Defensive statistics of the NBA Eastern Conference teams (TOV: turnovers of the opponents, PTS.Opp: points scored by the opponents). Grey line: TOV.O (opponents' turnovers; right vertical axis)

The Milwaukee Bucks are the team with the best performance on these defensive statistics (especially on defensive rebounds), while the Cleveland Cavaliers had the worst performance in defending (they had the lowest levels of steals, blocks, and rebounds). However, there is no evidence of a relationship between the defensive statistics and the points scored by the team's opponents or their turnovers. For example, the Atlanta Hawks (the leftmost team) suffered the highest number of scored points, although they have the highest number of opponents' turnovers (grey line) and do not show the worst defensive performance in terms of steals, blocks and rebounds (it is not the team with lowest bars).

This response should be integrated with other statistics related to the defensive performance, which is indirectly measured by the offensive performance of the opponents (for example, the points scored by the opponent teams). A possible solution is the examination of the teams' Four Factors, as done in the previous Question Q1.

Q3: Can we compare the game profiles of the three best centers in NBA 2018/19?

According to some basketball experts, the three best centres in the NBA season 2018/19 have been Anthony Davis (New Orleans Pelicans), Joel Embiid (Philadelphia 76ers) and Karl-Anthony Towns (Minnesota Timberwolves). We can compare their performance according to 2- and 3-point shots made (P2M and P3M), free throws made (FTM), total (offensive and defensive) rebounds (REB), assists (AST), steals (STL), and blocks (BLK) (all per minute played) by constructing three profile plots with the function `radialprofile`. Figure 2.3 shows that the profiles of the three considered players are very similar. Special attention must be paid in interpreting radial plots because the axes all have the same scale and sometimes this prevents us from seeing differences among players.

Figure 2.3: Radial plots of three selected centres, non-standardised variables. Dashed blue line: midpoint between minimum and maximum.

Indeed, the comparison should be complemented by analysing Figure 2.4, where variables have been standardised [3]. Here, the points in each profile show whether that player is positioned above or below the average (computed on the three analysed players) for the considered variables. Variable standardisation makes it possible to highlight differences among the three players. For example, focusing on the 3-point field goals, it is evident that Karl-Anthony Towns had the best performance, far above average, followed by Joel Embiid (close to

---

[3] A standardised variable is a variable that has been rescaled to have a mean of 0 and a variance (or a standard deviation) of 1. In order to standardise a variable, it is necessary to subtract the mean from each of its observed values and divide by the standard deviation. Standardising allows to compare variables, even when measured on different scales.

average) and then Anthony Davis (below average). Anthony Davis outperformed the two other centres in several other variables (P2M, BLK, STL, AST), while Joel Embiid is the best player (among the three) in free throws and rebounding.

Figure 2.4: Radial plots of three selected centres, standardised variables. Dashed blue line: zero (average of each variable).

Q4: Which team had the best shooting performance in NBA 2018/19?

Shooting performance can be measured by the success percentage of 2-point shots, 3-point shots, and free throws. The number of attempted shots is also important because it is different if a team's players have made all the shots (success percentage equal to 100) in 10 or 100 attempts. All these four variables can be represented for the NBA teams in one single plot using the function `bubbleplot`. Figure 2.5 shows every team as a bubble, whose location reflects its scoring percentages in 2-point shots (x-axis) and 3-point shots (y-axis), coloured according to the scoring percentage of free throws (on the red-blue colour scale) and sized according to the number of attempted shots (of any kind), rescaled between 0 and 100. Vertical and horizontal black lines indicate the average scoring percentages in 2-point shots and 3-point shots, respectively, computed over the teams considered in the plot.

Figure 2.5 highlights some outstanding teams: The Golden State Warriors show very high shooting percentages with a medium-low number of attempted shots. The San Antonio Spurs and the LA Clippers have similar performance on 2- and 3-point shots (exceptionally good in 3-point shots and just below the average in 2-point shots), but the Spurs exhibit a higher percentage of free throws made and a lower number of attempted shots. The Toronto Raptors, the NBA 2018/2019 champions, have above-average performance on all three

2.4. STATISTICS ANSWERS FANS' QUESTIONS

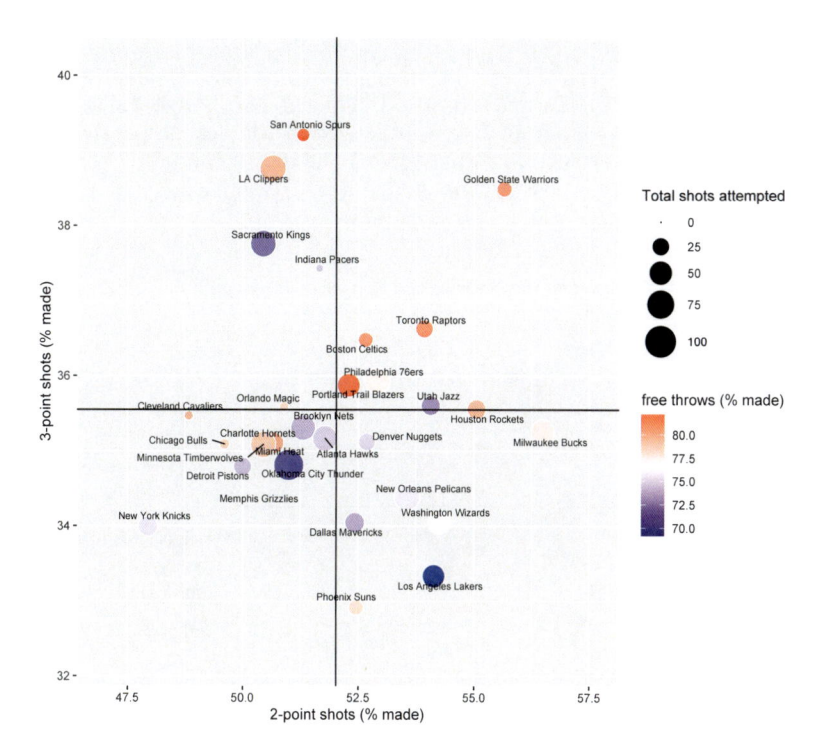

Figure 2.5: Bubble plot of teams according to the scoring percentages of 2-point shots (x-axis), 3-point shots (y-axis), and free throws (red-blue colour scale); bubble size: number of attempted shots.

types of shot but do not differ sharply from other teams (for example, Boston Celtics). The Portland Trail Blazers, finalists of the Western Conference together with the Golden State Warriors, excel at free throws but are close to the average for 2- and 3-point shots performance, with a high number of attempted shots.

Q5: Which players of the two NBA finalists had the best shooting performance and the best defence in NBA 2018/19?

A bubble plot analogous to the one in Figure 2.5 can be created to represent players instead of teams. Figures 2.6 and 2.7 show, respectively, the shooting and the defence performance of the players of the two NBA finalists (Golden State Warriors and Toronto Raptors) who have played at least 1,000 minutes in the regular season. Names are coloured to distinguish the team they belong to (red for Golden State Warriors and blue for Toronto Raptors). Vertical and horizontal black lines indicate the average of x-axis and y-axis variables, respectively, computed over the selection of players considered in the plot.

It is interesting to observe that in both plots there is a mix of players from the two teams in all the four quadrants.

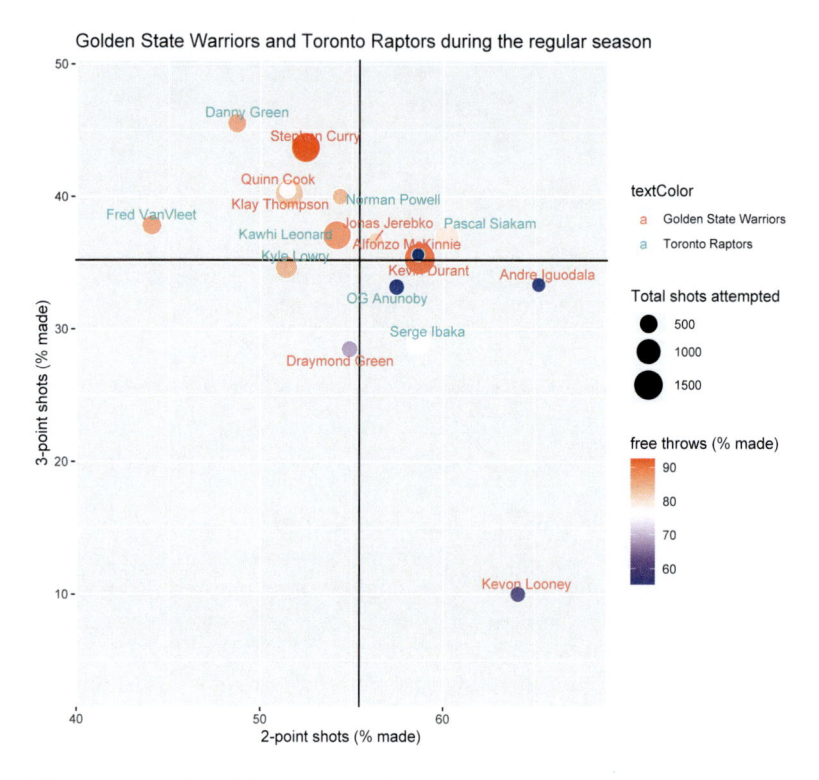

Figure 2.6: Bubble plot of selected players according to the scoring percentages of 2-point shots (x-axis), 3-point shots (y-axis), free throws (red-blue colour scale); bubble size: number of attempted shots.

If we must choose the best player from the shooting point of view, Figure 2.6 suggests Stephen Curry for the Golden State Warriors and Danny Green for the Toronto Raptors. The Raptors have relied heavily on Kawhi Leonard (one of the best defenders in NBA), whose position however is not as good as that of Danny Green, at least

2.4. STATISTICS ANSWERS FANS' QUESTIONS

regarding the analysed variables.

Kawhi Leonard appears in his great defensive ability in Figure 2.7, together with Draymond Green of Golden State Warriors, who shows, with respect to Leonard, a better performance on blocks but a lower number of steals (per minute played). Serge Ibaka shows a great performance in blocks: this allowed him to help the Raptors defeat the Golden State Warriors during the NBA Finals. It is useful to compare this plot with one of the top-10 lists of the best defenders released by NBA, which includes Klay Thompson, Draymond Green and Kawhi Leonard. Some fans were surprised by the exclusion of some players from this list, for example, Pascal Siakam and Danny Green. According to Figure 2.7, while Draymond Green and Kawhi Leonard have outstanding performance in defence, Klay Thompson has a position comparable to that of Danny Green, while Pascal Siakam even performed slightly better than Klay Thomson, especially with respect to blocks and defensive rebounds. A bubble plot using other interesting game variables can give another perspective for evaluating players.

Q6: How different is the performance on the 3-point shots among the players of Los Angeles Lakers?

Selecting only the players of Los Angeles Lakers who played at least 500 minutes and have attempted more than one 3-point shot, we have 11 players with different performances, as shown in Table 2.1.

Table 2.1: 3-point shots percentage (P3p) and 3-point shots attempted (P3A), Los Angeles Lakers

Player	P3p	P3A
Alex Caruso	48.00	50
Lance Stephenson	37.06	197
Rajon Rondo	35.92	142
Kentavious Caldwell-Pope	34.71	435
Reggie Bullock	34.34	99
LeBron James	33.94	327
Josh Hart	33.58	274
Brandon Ingram	32.98	94
Lonzo Ball	32.89	228
Kyle Kuzma	30.33	422
JaVale McGee	8.33	12

The average percentage is 32.92, but there is a great difference between the top-player Alex Caruso (48%) and JaVale McGee (8.33%).

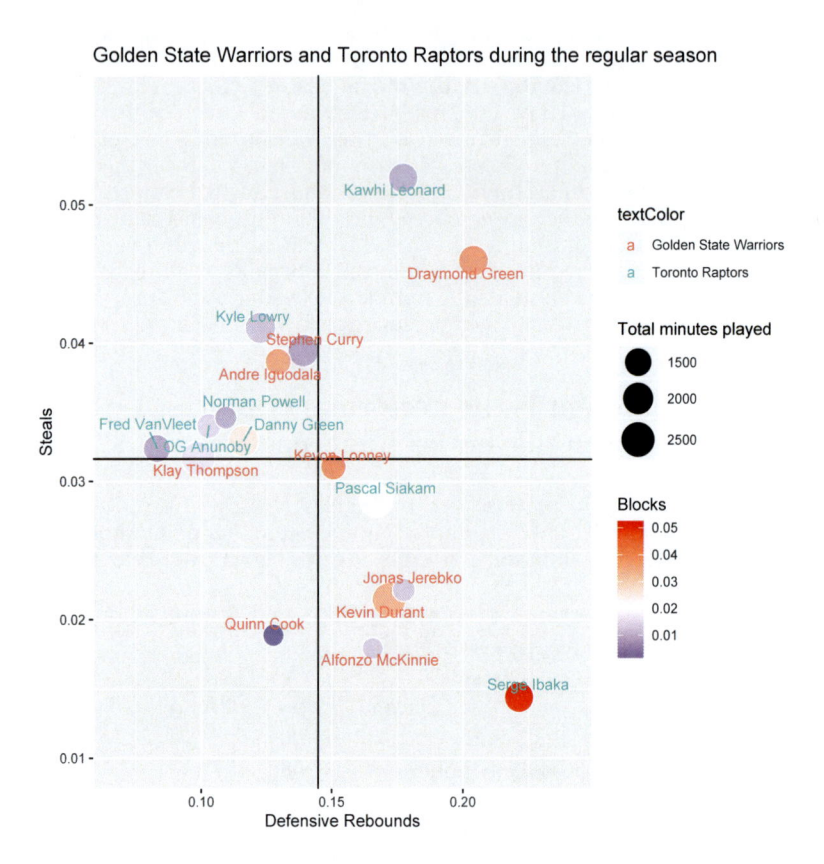

Figure 2.7: Bubble plot of selected players according to the defensive rebounds per minute played (x-axis), steals per minute played (y-axis), blocks per minute played (red-blue colour scale); bubble size: total number of minutes played.

2.4. STATISTICS ANSWERS FANS' QUESTIONS

To measure the variability of the 3-point shot scoring percentages, it is possible to compute some indexes with the function `variability`. Standard deviation, variation coefficient and range are equal to 8.90, 0.27 and 39.67, respectively.

An important feature to consider is the number of attempted shots when computing the variability indexes. Figure 2.8 shows a bubble for every player, located in correspondence of his 3-point percentage (vertical axis) with size proportional to the number of 3-point shots attempted. It is evident that there is a player with an outstanding position (Alex Caruso) with a relatively low number of attempted shots. Considering only the remaining players, the bubbles are not very scattered around the average, denoting a fairly low variability.

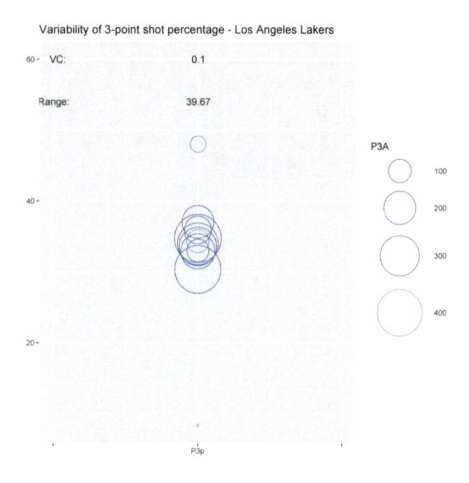

Figure 2.8: Variability diagram of the 3-point shot percentages weighted by the attempted shots, Los Angeles Lakers, VC=Variation Coefficient.

Q7: How balanced is the San Antonio Spurs 3-point shot performance?

This question could be addressed analysing the team variability as in the previous question Q6. However, if we want to understand if the San Antonio Spurs are a well-balanced team (that is, all the players contribute to its 3-point shot performance) or, on the contrary, depend too much on a few players, we can also consider the methods of inequality analysis, usually performed to evaluate the income or wealth distribution in a country. In our context, we can resort to inequality analysis to measure the distribution of game achievements within the players of a team. With the function `inequality`, we obtain the value for the Gini coefficient, which measures the degree of inequality ranging from 0% (null inequality) to 100% (maximum inequality). In the example, focusing on the performance of 3-point shots, we can study whether only one or a few players are able to score all the 3-point shots (high level of inequality) or, conversely, all the team players give an equal contribution (null inequality).

Considering the 10 players of San Antonio Spurs who have scored the highest number of 3-point shots (Table 2.2), the value of the Gini coefficient equals 39.38%. This denotes a quite high level of inequality, given that the Gini coefficient computed on the 2018-19 NBA data ranges from 17.71% (Boston Celtics) to 48.13% (Golden State Warriors), with average equal to 29.65%. The San Antonio Spurs can count on a few players that score much of the 3-point shots, so it is not well-balanced from this perspective. Indeed, the top three players (Forbes, Mills and Belinelli) scored 61% of the total number of 3-point shots scored by all 8 players considered.

Table 2.2: 3-point shots made (P3M) and 3-point shots attempted (P3A), San Antonio Spurs

Player	P3M	P3A
Bryn Forbes	176	413
Patty Mills	159	404
Marco Belinelli	147	395
Davis Bertans	145	338
Rudy Gay	74	184
Derrick White	48	142
Dante Cunningham	30	65
LaMarcus Aldridge	10	42

Figure 2.9 displays all 30 NBA teams according to their Gini coefficient's value and number of 3-point shots made. The San Antonio Spurs have roughly the same number of 3-point shots made as the LA Clippers or the Phoenix Suns, but with a highest level of inequality. The Houston Rockets appear as an outlier, with a huge number of 3-point shots made and a fairly high value of Gini coefficient: James Harden alone scored nearly 1/3 of the total number of 3-point shots made in the whole season.

2.4. STATISTICS ANSWERS FANS' QUESTIONS

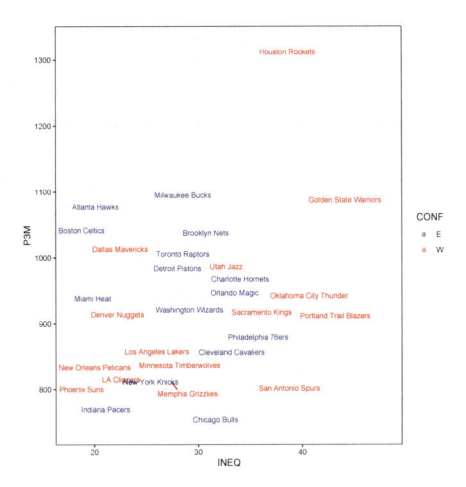

Figure 2.9: Gini coefficient (INEQ, *x*-axis) and number of 3-point shots made (P3M, *y*-axis) in the Western (red) and Eastern (blue) Conference teams.

Q8: What are LeBron James's favourite and disliked spots on the court?

LeBron James is one of the top NBA players; in 2017/2018 "King James" played with Cleveland Cavaliers before moving to Los Angeles Lakers in 2018. To identify his favourite and disliked spots on the court, the function `shotchart`, applied to play-by-play data (with space coordinates), allows to obtain interesting shot charts, such as those in Figure 2.10. In the left plot, points represent the shots attempted by LeBron James, coloured according to whether he missed the shot (blue) or scored the basket (red). In the right plot, the court is split into sectors coloured according to the average play length, that is the average time elapsed since the immediately preceding event when the shot is attempted. We note that in the first seconds of the play, LeBron James prefers shooting from short distances (especially from his left-hand side) or attempting long-range shots from the centre (where he has 40% of successful shots). Late shots are mainly attempted from the left and right sides (both mid-long and close range, with success rates ranging from 33% to 40%) and the centre mid-range (where the proportion of shots that scored the basket is 47%). LeBron James shoots from short range, on average, very early in the play, and these shots are the most successful (75%).

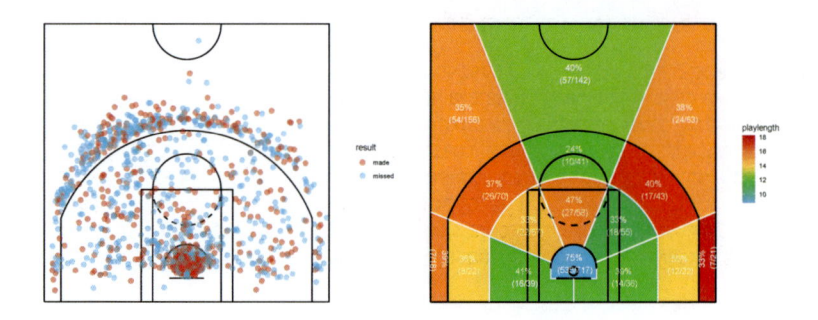

Figure 2.10: Shot chart (LeBron James). Left: missed (blue) and made (red) shots; Right: areas coloured according to the average play length and annotated with shooting statistics.

Q9: How does the network of assists work in the team of Cleveland Cavaliers?

The analysis of passing sequence with reference to assists (the last pass before shot) can be evaluated using network analysis tools. In BasketballAnalyzeR this is implemented in the function assistnet, which requires play-by-play data. Investigation of interactions among teammates is especially important in basketball analytics due to its nature of team sport.

The left-hand plot in Figure 2.11 displays the network of assists made and received by the Cleveland Cavaliers players. Each node represents a player and the oriented edge goes from the player who made the assist to the player who received it. The edges are coloured according to the number of assists. In the right plot, nodes are coloured according to the points scored thanks to a teammate's assist (FG-PTS_AST) and sized according to the points scored by assisted teammates (ASTPTS).

The crucial role played by LeBron James clearly emerges. He appears as the centre of the game strategy, offering lots of assists to his teammates, firstly to Kevin Love and, secondly, to Jeff Green, Kyle Korver (who also receives assists from Dwyane Wade), JR Smith, and Jae Crowder.

Node size and colour in the right plot in Figure 2.11 give interesting insights on the network of assists of Cleveland Cavaliers. For

2.4. STATISTICS ANSWERS FANS' QUESTIONS

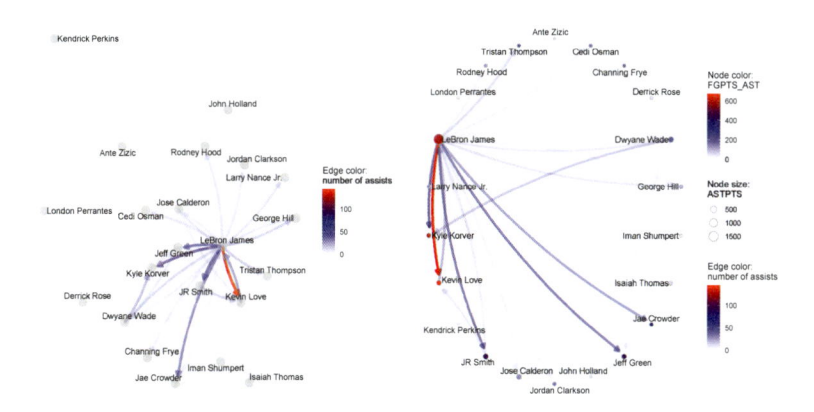

Figure 2.11: Network of assists of Cleveland Cavaliers. Edges are coloured according to the number of assists (see the white-blue-red scale in the legend). Right plot: nodes are coloured according to the points scored thanks to a teammate's assist (FGPTS_AST); node size is according to points scored by assisted teammates (ASTPTS).

example, LeBron James scored a high number of points thanks to a teammate's assist (his node is red), but he was also able to create scoring opportunities for the other players (his node is big). Kevin Love scored points thanks to the assists he received, while not offering many successful assists to his teammates. JR Smith, Jeff Green and Jae Cowder have small blue circles, indicating that they capitalised a medium-low number of assists and, at the same time, offered few fruitful assists to their teammates.

Q10: How does the frequency of 2-point shots by Cleveland Cavaliers vary during the match?

The analysis of the shooting frequency in time (or even in space) is a truly relevant topic in basketball analytics and requires play-by-play data. The function `densityplot` originates nice plots, like those in Figure 2.12, displaying the density estimation of the 2-point shots attempted by the players of Cleveland Cavaliers, with respect to two concurrent variables: the period time (left plot), i.e., the time played in the quarter (in seconds) and the total time (right plot), i.e., the time played in the match (in seconds). This analysis is made possible thanks to the availability, in the play-by-play data, of total time and play time for every single shot.

The Cleveland Cavaliers tend to concentrate their shots in the first

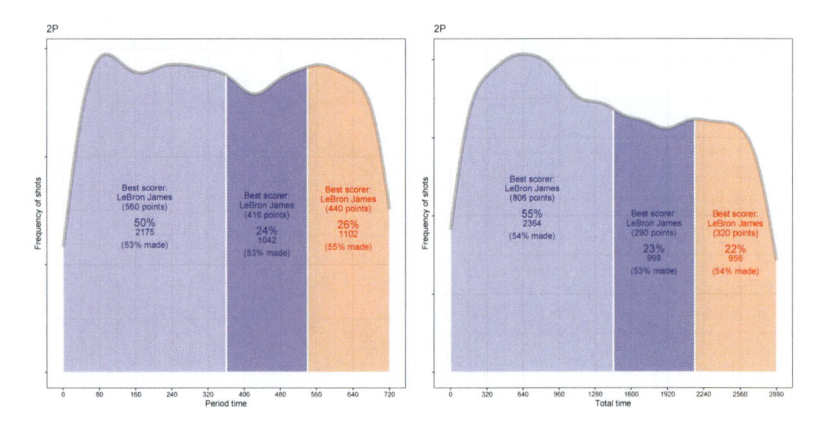

Figure 2.12: Density estimation of the 2-point shots by Cleveland Cavaliers with respect to period time (left) and total time (right).

half of the match (55%). Looking at the period time, they tend to equally divide their 2-point shots in the two halves of each quarter, with a slightly higher concentration in the last part (26%) than in the third one (24%). The team's success percentages are stable (53%-55%) during the different phases of the match or of each quarter. In every sub-period of time, LeBron James is always the best scorer (in brackets, the total number of points he scored). Several other interesting observations can be drawn comparing these results with those of the opponents of Cleveland Cavaliers or by focusing on some specific players.

Q11: Which is the best distance to shoot from for LeBron James? And for his teammates?

With the function `expectedpts` we can estimate the expected points of one team or single players with respect to some variables, such as the shot distance or the total time played in the match. Focusing on the expected points rather than on the scoring probability allows to determine, for each player, his situations of maximum efficiency (for example, the best distance to shoot from), considering both the points scored and their scoring probabilities.

Figure 2.13 shows the expected points by LeBron James (left plot) and all the players of Cleveland Cavaliers who scored more than 500 points (right plot) in function of the shot distance. The maximum efficiency of LeBron James is for the distances where the red line is above the grey line (team average): when he shoots from a distance higher than 22 feet (i.e., in 3-point shots) he clearly overperforms in

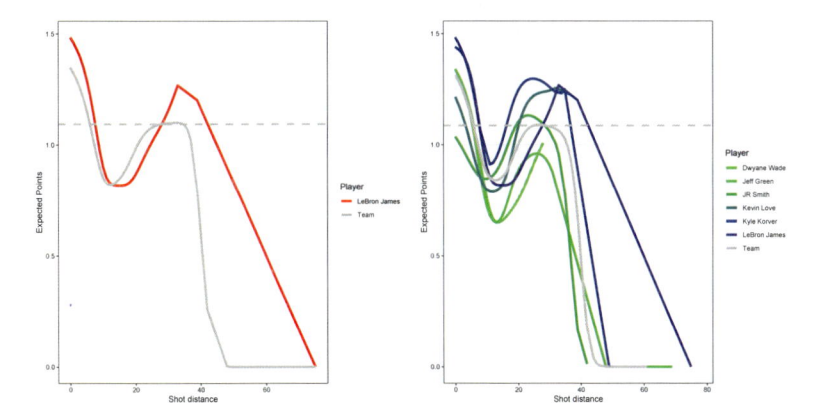

Figure 2.13: Expected points from a given distance (dashed line: team average independent from shot distance) of shots by LeBron James (left) and all the Cleveland Cavaliers players who scored more than 500 points (right).

terms of expected points. He performs better than the team average from every distance, except when shooting from a distance between 10 to 22 feet. From that distance, Kyle Korver, Kevin Love, and JR Smith perform better than LeBron James and the team average. This result can really help to identify the best players the team can count on in each spot of the court and, finally, to define a winning game strategy.

2.5 Conclusions

In this contribution the focus was on performance analysis in basketball in an era in which most coaches and their backroom staff rely on formulas and figures to predict the most effective methods for winning. This is especially true in the US, where NBA is leading the major transformation related to the use of analytics: experts rely on data to measure a team's probability of winning and to assess a player's or a team's performance. Assessments based on the "eye-test", that is the impression that came from watching a game, are out of fashion. The attention is in particular on basketball analytics used to analyse performance issues related to players and teams, playing patterns, game strategies, performance drivers, and interactions among players.

After briefly introducing the state of the art of basketball analytics and the different basketball data sets existing to perform statistical analyses, we offered an overview of possible basic and advanced investigations in performance analysis in basketball, with focus on some of the opportunities offered by the R package `BasketballAnalyzeR`. To show that statistics are more than numbers, we replied to several possible questions that could be asked by basketball fans, in a sort of Q&A section with questions from fans and answers by basketball analysts.

Basketball analytics can give the appropriate answers to many other questions. We believe that `BasketballAnalyzeR` can be a valid help for both basketball fans without a strong statistical background and expert analysts, who can subsequently apply their sophisticated methods to the results obtained from the proposed analyses.

The quality of the available data plays a crucial role in analytics. We focused on NBA data, which are complete and of reliable quality. Steps are also being taken in other countries, leagues and championships, in order to improve the data collection and finally obtain high-quality data, which is the ingredient needed to perform accurate analyses with reliable results that can be appreciated by sport professionals and fans. This can foster the spread of the statistical culture and finally refine the understanding of basketball analytics in a virtuous circle that is good for both basketball and statistics.

Acknowledgments

The Authors wish to thank BigDataBall (www.bigdataball.com) for making data available.

Bibliography

Aglioti, S. M., Cesari, P., Romani, M., and Urgesi, C. (2008). Action anticipation and motor resonance in elite basketball players. *Nature Neuroscience*, 11(9):1109–1116.

Alagappan, M. (2012). From 5 to 13: Redefining the positions in basketball. *MIT Sloan Sports Analytics Conference*.

Alamar, B. C. (2013). *Sports analytics: A Guide for Coaches, Managers, and Other Decision Makers*. Columbia University Press, New York.

Albert, J., Glickman, M. E., Swartz, T. B., and Koning, R. H. (2017). *Handbook of Statistical Methods and Analyses in Sports*. CRC Press, Boca Raton, Florida.

Annis, D. H. (2006). Optimal end-game strategy in basketball. *Journal of Quantitative Analysis in Sports*, 2(2):1–1.

Ante, P., Slavko, T., and Igor, J. (2014). Interdependencies between defence and offence in basketball. *Sport Science*, 7(2):62–66.

Arkes, J. (2010). Revisiting the hot hand theory with free throw data in a multivariate framework. *Journal of Quantitative Analysis in Sports*, 6(1):1–12.

Avugos, S., Köppen, J., Czienskowski, U., Raab, M., and Bar-Eli, M. (2013). The "hot hand" reconsidered: A meta-analytic approach. *Psychology of Sport and Exercise*, 14(1):21–27.

Bar-Eli, M., Avugos, S., and Raab, M. (2006). Twenty years of "hot hand" research: Review and critique. *Psychology of Sport and Exercise*, 7(6):525–553.

Barber, J. M. and Rollins, E. S. (2019). Factors influencing scoring in the NBA slam dunk contest. *Journal of Sports Analytics*, 5(2):223–245.

Bianchi, F., Facchinetti, T., and Zuccolotto, P. (2017). Role revolution: towards a new meaning of positions in basketball. *Electronic Journal of Applied Statistical Analysis*, 10(3):712–734.

Bornn, L., Cervone, D., Franks, A., and Miller, A. (2017). Studying basketball through the lens of player tracking data. In *Handbook of Statistical Methods and Analyses in Sports*, pages 245–269. Chapman and Hall/CRC, Boca Raton, Florida.

Bourbousson, J., Sève, C., and McGarry, T. (2010a). Space–time coordination dynamics in basketball: Part 1. Intra-and inter-couplings among player dyads. *Journal of Sports Sciences*, 28(3):339–347.

Bourbousson, J., Sève, C., and McGarry, T. (2010b). Space–time coordination dynamics in basketball: Part 2. The interaction between the two teams. *Journal of Sports Sciences*, 28(3):349–358.

Brown, M. and Sokol, J. (2010). An improved LRMC method for NCAA basketball prediction. *Journal of Quantitative Analysis in Sports*, 6(3):1–23.

Clemente, F. M., Martins, F. M. L., Kalamaras, D., and Mendes, R. S. (2015). Network analysis in basketball: inspecting the prominent players using centrality metrics. *Journal of Physical Education and Sport*, 15(2):212–217.

Cooper, W. W., Ruiz, J. L., and Sirvent, I. (2009). Selecting non-zero weights to evaluate effectiveness of basketball players with DEA. *European Journal of Operational Research*, 195(2):563–574.

Csataljay, G., O'Donoghue, P., Hughes, M., and Dancs, H. (2009). Performance indicators that distinguish winning and losing teams in basketball. *International Journal of Performance Analysis in Sport*, 9(1):60–66.

de Oliveira, R. F., Oudejans, R. R., and Beek, P. J. (2006). Late information pick-up is preferred in basketball jump shooting. *Journal of Sport Sciences*, 24:933–940.

De Rose, D. J. (2004). Statistical analysis of basketball performance indicators according to home/away games and winning and losing teams. *Journal of Human Movement Studies*, 47:327–336.

Deshpande, S. K. and Jensen, S. T. (2016). Estimating an NBA player's impact on his team's chances of winning. *Journal of Quantitative Analysis in Sports*, 12(2):51–72.

Engelmann, J. (2017). Possession-based player performance analysis in basketball (adjusted +/− and related concepts). In *Handbook of Statistical Methods and Analyses in Sports*, pages 215–227. Chapman and Hall/CRC.

Erčulj, F. and Štrumbelj, E. (2015). Basketball shot types and shot success in different levels of competitive basketball. *PloS one*, 10(6):e0128885.

Fearnhead, P. and Taylor, B. M. (2011). On estimating the ability of NBA players. *Journal of Quantitative Analysis in Sports*, 7(3):11–11.

Fewell, J. H., Armbruster, D., Ingraham, J., Petersen, A., and Waters, J. S. (2012). Basketball teams as strategic networks. *PloS one*, 7(11):e47445.

Franks, A. M., D'Amour, A., Cervone, D., and Bornn, L. (2016). Meta-analytics: tools for understanding the statistical properties of sports metrics. *Journal of Quantitative Analysis in Sports*, 12(4):151–165.

Fujimura, A. and Sugihara, K. (2005). Geometric analysis and quantitative evaluation of sport teamwork. *Systems and Computers in Japan*, 36(6):49–58.

García, J., Ibáñez, S. J., De Santos, R. M., Leite, N., and Sampaio, J. (2013). Identifying basketball performance indicators in regular season and playoff games. *Journal of Human Kinetics*, 36(1):161–168.

Gilovich, T., Vallone, R., and Tversky, A. (1985). The hot hand in basketball: On the misperception of random sequences. *Cognitive Psychology*, 17(3):295–314.

Goldman, M. and Rao, J. M. (2012). Effort vs. concentration: the asymmetric impact of pressure on NBA performance. *MIT Sloan Sports Analytics Conference*.

Gómez Sánchez, J., Moll Sotomayor, J. A., and Pila Teleña, A. (1980). Baloncesto: Técnica de entrenamiento y dirección de equipo. *Madrid, Pila Teleña*.

Groll, A., Manisera, M., Schauberger, G., and Zuccolotto, P. (2018). Guest Editorial 'Statistical modelling for sports analytics'. *Statistical Modelling*, 18(5-6):385–387.

Groll, A., Manisera, M., Schauberger, G., and Zuccolotto, P. (2019). Guest Editorial 'Statistical modelling for sports analytics'. *Statistical Modelling*, 19(1):3–4.

Gudmundsson, J. and Horton, M. (2017). Spatio-temporal analysis of team sports. *ACM Computing Surveys*, 50(2).

Gupta, A. A. (2015). A new approach to bracket prediction in the NCAA men's basketball tournament based on a dual-proportion likelihood. *Journal of Quantitative Analysis in Sports*, 11(1):53–67.

Ibáñez, S. J., García, J., Feu, S., Lorenzo, A., and Sampaio, J. (2009). Effects of consecutive basketball games on the game-related statistics that discriminate winner and losing teams. *Journal of Sports Science and Medicine*, 8(3):458–462.

Ibáñez, S. J., Sampaio, J., Sáenz-López Buñuel, P., Giménez Fuentes-Guerra, J., and Janeira, M. (2003). Game statistics discriminating the final outcome of junior world basketball championship matches (Portugal 1999). *Journal of Human Movement Studies*, 45(1):1–20.

Kay, H. K. (1966). *A statistical analysis of the profile technique for the evaluation of competitive basketball performance*. PhD thesis, University of Alberta.

Kenett, R. S. and Redman, T. C. (2019). *The Real Work of Data Science: Turning data into information, better decisions, and stronger organizations*. John Wiley & Sons, Hoboken, New Jersey.

Koehler, J. J. and Conley, C. A. (2003). The "hot hand" myth in professional basketball. *Journal of Sport and Exercise Psychology*, 25(2):253–259.

Koh, K. T., Wang, C. K. J., and Mallett, C. (2011). Discriminating factors between successful and unsuccessful teams: A case study in elite youth Olympic basketball games. *Journal of Quantitative Analysis in Sports*, 7(3):21–21.

Koh, K. T., Wang, C. K. J., and Mallett, C. (2012). Discriminating factors between successful and unsuccessful elite youth Olympic female basketball teams. *International Journal of Performance Analysis in Sport*, 12(1):119–131.

Kubatko, J., Oliver, D., Pelton, K., and Rosenbaum, D. T. (2007). A starting point for analyzing basketball statistics. *Journal of Quantitative Analysis in Sports*, 3(3):1–22.

Lamas, L., De Rose Jr., D., Santana, F. L., Rostaiser, E., Negretti, L., and Ugrinowitsch, C. (2011). Space creation dynamics in basketball offence: validation and evaluation of elite teams. *International Journal of Performance Analysis in Sport*, 11(1):71–84.

Loeffelholz, B., Bednar, E., and Bauer, K. W. (2009). Predicting NBA games using neural networks. *Journal of Quantitative Analysis in Sports*, 5(1):1–15.

Lopez, M. J. and Matthews, G. J. (2015). Building an NCAA men's basketball predictive model and quantifying its success. *Journal of Quantitative Analysis in Sports*, 11(1):5–12.

Lorenzo, A., Gómez, M. Á., Ortega, E., Ibáñez, S. J., and Sampaio, J. (2010). Game related statistics which discriminate between winning and losing under-16 male basketball games. *Journal of Sports Science & Medicine*, 9(4):664–668.

Madden, C. C., Kirkby, R. J., McDonald, D., Summers, J. J., Brown, D. F., and King, N. J. (1995). Stressful situations in competitive basketball. *Australian Psychologist*, 30(2):119–124.

Madden, C. C., Summers, J. J., and Brown, D. F. (1990). The influence of perceived stress on coping with competitive basketball. *International Journal of Sport Psychology*, 21(1):21–35.

Manisera, M., Sandri, M., and Zuccolotto, P. (2019). BasketballAnalyzeR: the R package for basketball analytics. In *Conference Smart Statistics for Smart Applications*, SIS 2019, pages 395–402, Milan, Italy. Pearson.

Manner, H. (2016). Modeling and forecasting the outcomes of NBA basketball games. *Journal of Quantitative Analysis in Sports*, 12(1):31–41.

McFarlane, P. (2019). Evaluating NBA end-of-game decision-making. *Journal of Sports Analytics*, 5(1):17–22.

Metulini, R., Manisera, M., and Zuccolotto, P. (2017a). Sensor analytics in basketball. In Francesco, C. D., Giovanni, L. D., Ferrante, M., Fonseca, G., Lisi, F., and Pontarollo, S., editors, *Proceedings of the 6th International Conference on Mathematics in Sport*, pages 265–274, Padova, Italy. Padova University Press.

Metulini, R., Manisera, M., and Zuccolotto, P. (2017b). Space-time analysis of movements in basketball using sensor data. In Petrucci, A. and Verde, R., editors, *Statistics and Data Science: new challenges, new generations - Proceedings of the SIS 2017 Conference of the Italian Statistical Society*, pages 701–706, Firenze, Italy. Firenze University Press.

Metulini, R., Manisera, M., and Zuccolotto, P. (2018). Modelling the dynamic pattern of surface area in basketball and its effects on team performance. *Journal of Quantitative Analysis in Sports*, 14(3):117–130.

Miller, A. C. and Bornn, L. (2017). Possession sketches: Mapping NBA strategies. MIT *Sloan Sports Analytics Conference 2017*.

Miller, S. and Bartlett, R. (1996). The relationship between basketball shooting kinematics, distance and playing. *Journal of Sports Sciences*, 14(3):243–253.

Miller, T. W. (2015). *Sports Analytics and Data Science: Winning the Game with Methods and Models*. FT Press.

Okubo, H. and Hubbard, M. (2006). Dynamics of the basketball shot with application to the free throw. *Journal of Sports Sciences*, 24(12):1303–1314.

Oliver, D. (2004). *Basketball on Paper: Rules and Tools for Performance Analysis*. Potomac Books, Inc.

Özmen, U. M. (2012). Foreign player quota, experience and efficiency of basketball players. *Journal of Quantitative Analysis in Sports*, 8(1):1–18.

Page, G. L., Barney, B. J., and McGuire, A. T. (2013). Effect of position, usage rate, and per game minutes played on NBA player production curves. *Journal of Quantitative Analysis in Sports*, 9(4):337–345.

Page, G. L., Fellingham, G. W., and Reese, S. C. (2007). Using boxscores to determine a position's contribution to winning basketball games. *Journal of Quantitative Analysis in Sports*, 3(4):1–1.

Passos, P., Araújo, D., and Volossovitch, A. (2016). *Performance Analysis in Team Sports*. Taylor & Francis.

Passos, P., Davids, K., Araújo, D., Paz, N., Minguéns, J., and Mendes, J. (2011). Networks as a novel tool for studying team ball sports as complex social systems. *Journal of Science and Medicine in Sport*, 14(2):170–176.

Perše, M., Kristan, M., Kovačič, S., Vučkovič, G., and Perš, J. (2009). A trajectory-based analysis of coordinated team activity in a basketball game. *Computer Vision and Image Understanding*, 113(5):612–621.

Piette, J., Anand, S., and Zhang, K. (2010). Scoring and shooting abilities of NBA players. *Journal of Quantitative Analysis in Sports*, 6(1):1–1.

Piette, J., Pham, L., and Anand, S. (2011). Evaluating basketball player performance via statistical network modeling. MIT *Sloan Sports Analytics Conference*.

Ruiz, F. J. and Perez-Cruz, F. (2015). A generative model for predicting outcomes in college basketball. *Journal of Quantitative Analysis in Sports*, 11(1):39–52.

Sampaio, J., Drinkwater, E. J., and Leite, N. M. (2010). Effects of season period, team quality, and playing time on basketball players' game-related statistics. *European Journal of Sport Science*, 10(2):141–149.

Sampaio, J. and Janeira, M. (2003). Statistical analyses of basketball team performance: understanding teams' wins and losses according to a different index of ball possessions. *International Journal of Performance Analysis in Sport*, 3(1):40–49.

Sandri, M. (2020). The R package BasketballAnalyzeR. In *Basketball Data Science (Chapter 6)*. Chapman and Hall/CRC.

Severini, T. A. (2014). *Analytic Methods in Sports: Using Mathematics and Statistics to Understand Data from Baseball, Football, Basketball, and Other Sports*. Chapman and Hall/CRC.

Shortridge, A., Goldsberry, K., and Adams, M. (2014). Creating space to shoot: quantifying spatial relative field goal efficiency in basketball. *Journal of Quantitative Analysis in Sports*, 10(3):303–313.

Silverman, B. W. (1986). *Density Estimation for Statistics and Data Analysis*. Chapman and Hall/CRC.

Skinner, B. (2010). The price of anarchy in basketball. *Journal of Quantitative Analysis in Sports*, 6(1):3–3.

Skinner, B. and Goldman, M. (2017). Optimal strategy in basketball. In *Handbook of Statistical Methods and Analyses in Sports*, pages 229–244. Chapman and Hall/CRC.

Therón, R. and Casares, L. (2010). Visual analysis of time-motion in basketball games. In *Smart Graphics*, pages 196–207. Springer.

Travassos, B., Araújo, D., Davids, K., Esteves, P. T., and Fernandes, O. (2012). Improving passing actions in team sports by developing interpersonal interactions between players. *International Journal of Sports Science & Coaching*, 7(4):677–688.

Trninić, S., Dizdar, D., and Lukšić, E. (2002). Differences between winning and defeated top quality basketball teams in final tournaments of European club championship. *Collegium Antropologicum*, 26(2):521–531.

Tversky, A. and Gilovich, T. (2005). The cold facts about the "hot hand" in basketball. *Anthology of Statistics in Sports*, 16:169.

Vergin, R. (2000). Winning streaks in sports and the misperception of momentum. *Journal of Sport Behavior*, 23(2):181–197.

Vračar, P., Štrumbelj, E., and Kononenko, I. (2016). Modeling basketball play-by-play data. *Expert Systems with Applications*, 44:58–66.

West, B. T. (2008). A simple and flexible rating method for predicting success in the NCAA basketball tournament: Updated results from 2007. *Journal of Quantitative Analysis in Sports*, 4(2):1–18.

Winston, W. L. (2012). *Mathletics: How Gamblers, Managers, and Sports Enthusiasts use Mathematics in Baseball, Basketball, and Football*. Princeton University Press.

Wu, S. and Bornn, L. (2018). Modeling offensive player movement in professional basketball. *The American Statistician*, 72(1):72–79.

Yuan, L.-H., Liu, A., Yeh, A., Kaufman, A., Reece, A., Bull, P., Franks, A., Wang, S., Illushin, D., and Bornn, L. (2015). A mixture-of-modelers approach to forecasting NCAA tournament outcomes. *Journal of Quantitative Analysis in Sports*, 11(1):13–27.

Zhang, T., Hu, G., and Liao, Q. (2013). Analysis of offense tactics of basketball games using link prediction. In *Computer and Information Science (ICIS), IEEE/ACIS 12th International Conference*, pages 207–212. IEEE.

Zuccolotto, P. and Manisera, M. (2020). *Basketball Data Science: With Applications in R*. Chapman and Hall/CRC.

Zuccolotto, P., Manisera, M., and Kenett, R. (2017). Guest Editorial 'Statistics in sports'. *Electronic Journal of Applied Statistical Analysis*, 10(3):1–2.

Zuccolotto, P., Manisera, M., and Sandri, M. (2018). Big data analytics for modeling scoring probability in basketball: The effect of shooting under high-pressure conditions. *International Journal of Sports Science & Coaching*, 13(4):569–589.

Zuccolotto, P., Sandri, M., and Manisera, M. (2019). Spatial performance indicators and graphs in basketball. *Social Indicators Research*. In Press, https://doi.org/10.1007/s11205-019-02237-2.

Chapter 3
Measurement, Meaning, and Prediction in Sports

Thomas W. Miller
Northwestern University

Abstract

We provide a review of traditional measurement theory and practice in sports, highlighting desirable properties of measures. We show how to use traditional unidimensional scaling to assess the relative performance of teams with disparate schedules. We suggest entropy as a descriptor of player repertoire, and we introduce neural network embeddings as an alternative to a traditional measurement approach. Examples from baseball and basketball illustrate measurement methods.

3.1 Introduction—Words and Numbers

Words are central to thinking, measurement is central to science, and the two go hand-in-hand. Words describe attributes. And measurement, as commonly defined, is the assignment of numbers to attributes according to rules. Science meets sports when we measure on the fields and courts of play.

Many factors contribute to success in sport. There are measures that relate to physical stature, biophysics, health, fitness, and conditioning. There is athleticism and measures dealing with speed, power, strength, flexibility, and agility. We also have psychological measures of intelligence, personality, and attitude; and there is knowledge of the game. Proficiency in sport—demonstrated skill and execution in practice and games—is the objective of most athletes and teams. Sports measurement, analytics, and data science methods help athletes and teams to attain proficiency in sport.

In Section 2 we begin by reviewing measurement best practices. Measurement of athletic proficiency as it relates to individual players is covered in Section 3. We provide perspective on the levels of measurement controversy in Section 4, followed by a discussion of team rankings and ratings in Section 5. Innovative approaches to measurement are introduced in Section 6 on entropy and Section 7 on neural network embeddings. Final sections provide an overview of measurement theory and methods. Examples focus on baseball and basketball, although most reviewed concepts and methods apply to all sports.

Throughout the discussion, it will be useful to think of the role of words and numbers. Traditional measurement begins with words and ends with numbers. The embeddings approach begins and often ends with numbers.

3.2 Measurement Best Practices

We prefer sports measurements with certain desirable properties (Miller, 2016):

- **Reliable.** A measure should be trustworthy and repeatable.

- **Valid.** A measure should measure the attribute it is said to measure.

- **Explicit.** A measurement procedure should be unambiguous and defined in detail so that each research worker obtains the same values when using the procedure.

- **Accessible.** A measure should come from data that are easily obtained.

3.2. MEASUREMENT BEST PRACTICES

- **Tractable.** A measure should be easy to work with and easy to utilise in methods and models.
- **Comprehensible.** A measure should be simple and straightforward, so it is easily understood and interpreted.
- **Transparent.** The method of measurement should be documented fully so research workers can share results with one another in a spirit of open and honest scientific inquiry. There should be no trade secrets in science.

We use the term *reliability* to refer to the trustworthiness or repeatability of measurement procedures. We consider the degree to which repeated measures of the same attribute at the same time agree with one another, as in test-retest reliability or split-half reliability. When using a multi-item survey, we ask that a measure have internal consistency.

Reliability of many sports' performance measures is high because they are objective measures based on counts, the official scoring of events on the field, as reported in box scores and play-by-play logs. Official records, counts, and mathematical formulas for computing performance measures do not change from one observer or analyst to the next. Furthermore, many contemporary measures of player and ball location on the fields and courts of play, player running speed, and efficiency in getting to balls in play are obtained through electronic monitoring with little or no human intervention. These are highly reliable and trustworthy.

Sports analysts sometimes confuse reliability with stability. In baseball, for example, analysts may observe variability in player batting averages or earned run averages from one year to the next, saying that this is evidence of low reliability. This is incorrect thinking. Reliability concerns reproducibility of metrics, not stability over time.

When we talk about *validity*, we are thinking of the degree to which a measure measures what it is said to measure. Measures differ widely in validity, which we review in Section 8.

Sports performance measures also vary widely in the degree to which they possess attributes other than reliability and validity. Individual performance measures in baseball serve to illustrate measurement principles.

Simple, comprehensible measures are preferred to complex measures. Simple measures are easier to explain to fans, coaches, and managers. Among the most comprehensible individual measures are percentages or proportions computed from the events of a game. Batting average (BA) is a proportion with at-bats as a divisor. We look for players with batting averages above 0.250, and batting at or above 0.300 is a goal of many hitters. Batting below 0.200, sometimes referred to as "the Mendoza line", is not a good sign for hitters. Batting average is easily understood but criticised as a general measure of hitting ability because it fails to consider the value of walks. On-base percentage (OBP) is easy to explain. Using plate appearances rather than at-bats in the divisor, OBP reflects the proportion of times that a hitter reaches first base or beyond. For OBP we look for players whose values are around 0.333, getting on base one in every three plate appearances. OBP is well known and well understood, partly as a result of it being discussed in *Moneyball* (Lewis, 2003) and Sabermetrics (Baumer and Zimbalist, 2014). Nevertheless, OBP is criticised because it fails to consider the value of extra-base hits, another aspect of hitting prowess.

Slugging percentage (SLG) is a ratio: the number of total bases per at-bat. Because it is neither a percentage nor an average, it can be difficult to understand. A fan must be "in the know" to understand that an SLG value of 0.300 is bad or a value of 0.500 is good. Babe Ruth's lifetime SLG was 0.690, which is particularly good. SLG suffers from being less comprehensible than other measures of hitting prowess. Also lacking in comprehensibility is on-base percentage plus slugging (OPS), the sum of OBP and SLG. It is neither a percentage like OBP, nor a ratio like SLG. The intent of OPS is to provide an index of ability that reflects both getting on base and hitting with power. OPS is another measure for which a person needs to be "in the know" to understand.

OPS makes little intuitive sense, although we have seen an upward trend in recent years. The average regular-season OPS across the thirty Major League Baseball (MLB) teams in the 2019 season was 0.758 (Sports Reference LLC, 2019b), up from 0.700 in 2014 (Sports Reference LLC, 2019a). There is little justification for weighting OBP and SLG equally in measuring hitting prowess and adding a proportion to a ratio gives a measure with unknown units. OPS is neither comprehensible nor mathematically appropriate.

Tango et al. (2007) suggest an alternative to OPS called the on-base average (OBA), computed as a combination of various hitting measures. Their intent is to define a measure that gives reasonable, data-based weights to getting on base and hitting with power, a measure that is then scaled so it has values to conform to OBP. Their index represents a weighted linear combination of the number of intentional bases on balls, the number of times a player is hit by a pitch, the number of times a batter reaches base on an error, and the numbers of singles, doubles, triples, and home runs. OBA cannot be easily explained in words, so it fails as a general index of hitting prowess.

Much time and effort has been devoted to attempts at finding the best single measure of ability for position players. A five-tool player in baseball is a player with strong skills for running, fielding, throwing, hitting, and hitting with power. How can we combine measures of these five traits into a single measure reflecting a player's contribution

3.2. MEASUREMENT BEST PRACTICES

to his team?

Comprehensive player evaluation is illustrated by points or wins above replacement. The general idea is to assess player abilities in hitting, fielding, base running, and throwing relative to a norm group (referred to as replacement players) and then to combine those norm-group-scaled assessments.

We ask, "What is a player's value to his team? What if he were replaced by another player who is available to play, a player of average ability at the same position?" Wins above replacement is usually expressed in units of wins across the regular season, with ten runs being equivalent to one win. If a player's wins-above-replacement value is 5, say, then that player's team can expect to win five fewer games across the entire season if he is replaced.

Wins-above-replacement measures fail the transparency test when methods of calculation are closely held company secrets. This is a special problem for norm-group-based measures because their meaning rests on the choice of norm group. If we do not know who the replacement players are, then we cannot accurately interpret wins-above-replacement. Furthermore, there is no way of checking the calculations of for-profit companies that refuse to publish their formulas and data. These measures are not in keeping with the spirit of scientific inquiry. They are neither comprehensible nor transparent. For a transparent wins-above-replacement method, we can turn to openWAR from Baumer et al. (2015). Data and programs for this metric are in the public domain. What about player performance over time? There are truisms in life, and one of those truisms is that the body ages. Much is understood about age effects in baseball and how to model them (Fair, 2008). PECOTA, another proprietary measurement and prediction system, uses player-comparable age curves as its base data (Silver, 2004, 2012).

Fortunately, there are alternatives to PECOTA. Teams desiring age-based measures can compute them directly, obtaining predictions about performance over the course of a player's career (Albert and Bennett, 2001; Marchi and Albert, 2014). Some methods build on Bayesian inference (Albert, 2009). Age-based models are most easily developed using tractable measures such as proportions. Individual measures of athletic performance, with the exception of physical and physiological measures (Martin, 2016) and field and track measures under controlled conditions, are intrinsically flawed because they fail to consider the context of play. They fail to account for when events occur within sporting contests, who is playing for the individual athlete's team, and who is playing for the opposing team. A critical consideration in the measurement of baseball hitting prowess, for example, is the pitcher-batter matchup, as we will explore later with the embeddings approach to measurement.

3.3 Individual Contribution to Team

How do we go beyond individual player performance to look at a player's contribution to his or her team? Therein lies a fundamental question in sports analytics. Anyone who knows sports knows that a good team is worth more than the sum of its parts. And it should come as no surprise that a dysfunctional team is worth less than the sum of its parts. This is to say that team effects should be considered when predicting winners and losers.

Baseball may be less susceptible to team effects than other sports. On a baseball diamond, Tinker-to-Evers-to-Chance works fine, even when Tinkers, Evers, and Chance are not speaking with one another. Baseball is distinct from many other team sports in being defined by many individual discrete events, easily identified as belonging to one player or to pairs of players in a pitcher-batter matchup.

By their very nature, football, basketball, hockey, and many other team sports present special problems in evaluating individual players. Unlike baseball, these sports involve continuous plan and situations in which players complement one another. Some players are described as "team players" because they help their teammates play better. Stockton and Malone worked together as a unit on the Utah Jazz. Their classic pick-and-roll made the Jazz a difficult team to beat for many years. Individual performance measures for Stockton and Malone are inextricably intertwined (Oliver, 2004).

Teammates affect individual player metrics as reported in box scores. Consider a two-on-two basketball game, ignoring characteristics of the opposing team for a moment. On this two-person team, Mary has the option of playing with Joan, an exceptional scorer, or with Helen, an exceptional passer. If Mary's objective is to embellish her own individual scoring statistics, then she would clearly choose to play with Helen, the exceptional passer. How shall Mary and Helen be judged in terms of their contribution to the team? What is the value of assists relative to points scored?

Straight-up, head-to-head situations are difficult to identify in team sports typified by long periods of continuous play. Plus-minus metrics, originally developed for hockey (Wikipedia, 2019), have been utilised extensively in basketball (Englemann, 2017). Plus-minus metrics in basketball are based on points scored for and against a team while a player is on the court. Their computation requires extensive, accurate play-by-play information. Plus-minus metrics represent a way of looking beyond individual scoring statistics to assess contribution to the team.

Many regard plus-minus metrics as more reflective of individual player performance than box score statistics. Described by Oliver (2004) as "the Holy Grail of player ratings", plus-minus metrics are among the best known indices of individual contribution to team. As Winston puts it, "Basketball is a team game. The definition of a good player is somebody who makes his team better, not a player who scores 40 points a game" (Winston (2009), 202).

A traditional, unadjusted plus-minus measure for a basketball player is computed as the points scored by the player's team minus the points scored by opponent teams when the player is on the court, normalised to adjust for the time of play. Plus-minus player ratings are computed from play-by-play data, noting the players on the court for both teams, scores at the start and end of each possession, and the proportion of game time associated with each possession.

As we might expect, a player's simple plus-minus rating is affected by the quality of the player's team and the quality of opposing teams. Most players on good teams have positive plus-minus ratings, whereas most players on bad teams have negative plus-minus ratings.

There are methods of adjusting plus-minus ratings. Oliver (2004) suggests simulating a player's behaviour across all possible combinations of teammates, all possible combinations of coaches, and all possible combinations of opposing teams. Winston (2009) suggests solving for sets of player ratings that, when aggregated within opposing teams, provide the closest match to final game scores for those teams. Englemann (2017) reviews many alternatives for adjusting plus-minus ratings.

Even after adjustment, a player's value on one team may be quite different from his value on another team. This is revealed in free-agency markets in which bids for players vary widely across bidding teams. It is understood that player trades can be beneficial for all teams involved.

3.4 Levels of Measurement Redux

No review of measurement theory would be complete without some mention of S. S. Stevens, who wrote *On the Theory of Scales of Measurement* in 1946 (Stevens, 1946). This influential article identified four types of measures or scales: nominal, ordinal, interval, and ratio.

Table 3.1 summarises scale types or levels of measurement from Stevens (1946). The formal definition of a scale follows from its mathematical properties and the set of data transformations that, when used on the original measures, will create new measures with the same mathematical properties as the original measures.

For nominal scales, any one-to-one transformation will preserve the number of categories, the scale's essential property. Like numbers on player jerseys, nominal measures are naming devices or identifiers. We often assign numbers to levels or classes of categorical variables.

These, too, are nominal measures.

For ordinal scales, any one-to-one monotonic transformation will preserve the property of order. Ordinal measures are ubiquitous in competitive sports. When one team beats another, we observe an ordered pair. Across many teams and games, we observe a set of paired comparisons, sometimes logically consistent, sometimes not. After team A beats team B and team B beats C, we might expect that team A will also beat team C. But team C sometimes beats team A, establishing a "circular triad", a logical inconsistency in the ordering of teams.

Deriving interval measures from ordinal data, including data with circularities or inconsistencies, is the work of scaling, a collection of psychometric methods (Torgerson, 1958). In the next section we employ unidimensional scaling to derive interval-level scores or ratings of basketball teams, given only the win-loss experiences of those teams. Any one-to-one linear transformation (function of the form $y = ax + b$, with x being the original measure and y the new measure) will preserve the properties of an interval scale.

When sports enthusiasts talk about sports statistics, they are usually referring to ratio measures. Ratio scales are similar to interval scales except that the zero point must be preserved. For ratio measures, a data transformation that preserves scale properties will have the form $y = ax$. Selected in-game events are associated with players or teams. Sports statistics, as reported in box scores, are event counts and ratios of event counts. These are ratio measures.

Researchers following Stevens' dictums constitute the weak measurement school (Baker et al., 1966). They argue that many measures are ordinal rather than interval and that statistics relying on sums or differences, including means and variances, would be inappropriate for ordinal measures. Researchers following the strong statistics school, on the other hand, argue that statistical methods make no explicit assumptions about the meaning of measurements or their relationships to underlying dimensions—strong statistics may be used with weak measurements (Baker et al., 1966).

Despite objections from weak measurement believers, we understand that there are many situations in which computing the mean of ranks makes perfectly good sense. Unidimensional scaling methods provide a rationale for converting ordinal measures to interval measures. Aggregate nominal data represent counts, and counts are ratio measures. Strict rules regarding permissible statistics belie practical considerations about which statistics are appropriate for particular research situations.

For practical purposes we ask whether or not a variable has meaningful magnitude. If a variable is categorical, it lacks meaningful magnitude. Also, regarding categorical variables, we note whether

Table 3.1: Levels of Measurement

Level of Measurement (Scale Type)	Basic Empirical Operations	Mathematical Group Structure	Examples of Permissible Statistics
Nominal	equality, numbers like names	one-to-one correspondence	number of cases in class, frequency table, modal class
Ordinal	greater than, less than	one-to-one monotonic	median, percentiles, rank-order correlation
Interval	equality of intervals	one-to-one linear	mean, standard deviation, product-moment correlation
Ratio	equality of ratios	one-to-one linear, preserving the zero point	same statistics as interval level

Source: Adapted from Stevens (1946).

the attribute being measured is binary (taking only two possible values) or multinomial (taking more than two possible values). If we can make these simple distinctions across measures, we can do much useful research, computing statistics appropriate for the tasks at hand.

3.5 Rating Teams with Disparate Schedules

The strength of a team is defined in relation to other teams. When picking winning teams, it is not sufficient to consider individual player statistics. We must see how teams compete with one another.

Sports team rankings and models for predicting sporting event outcomes have garnered considerable attention from mathematicians as well as sports enthusiasts. Ranking methods, the use of rankings in the design of sporting tournaments, and the seeding of teams in tournaments have been discussed extensively (Appleton, 1995; Carlin, 1996; Groeneveld, 1990; Thompson, 1975; West, 2006). Ranking methods for football have been reviewed by Ley et al. (2019); Van Eetvelde and Ley (2019). Langville and Meyer (2012) provide a comprehensive review of rating and ranking algorithms and methods.

There is debate about how to rank teams, especially when teams have limited opportunity to play one another. Most of us remember the extensive controversies surrounding the Bowl Championship

Series (BCS) for college football prior to the introduction of a limited playoff program. There still remains controversy about which teams should qualify for playoffs due to strength-of-schedule differences across teams. Compared with college athletics, professional team sports have more balance across league schedules, but perfect balance eludes even the best of planners and scheduling algorithms. Divisions and conferences are not equal in player abilities, and teams play more of their games with teams in their own conferences and divisions.

One method for rating teams based on wins and losses comes from work on chess player rankings by Árpád Élö (1903–1992). The Elo system has been adapted to professional team sports in the United States and is the basis of Jeff Sagarin's rankings in *USA Today*.

In ranking sports teams we can go beyond binary win/loss data to consider the margin of victory or game point differentials. Again, we must account for variability in team schedules. When looking at points scored in games, we must also adjust for games involving extra innings in baseball or overtime periods in football or basketball.

A consideration when working with scores relates to very high scoring games. These may result from games involving teams that are unevenly matched ("blowouts") or games that have little bearing on team possibilities for moving into the playoffs. In games such as these, teams may be using second string or bench players in place of their usual line-ups. Some adjustment of game scores would be appropriate in such instances.

Other approaches to team rankings involve developing separate measures or ranks for offensive and defensive performance, which are later combined into a single measure or rank. Sometimes it makes sense to employ a multivariate approach, accounting for team performance on many measures.

How shall we judge or rank teams, given imbalance in team schedules? One way is to employ unidimensional scaling from traditional psychometrics. We examine win/loss data from team matchups as paired comparisons, one team over another in each game.

Consider the problem of assessing team strength in the National Basketball Association (NBA). Table 3.2 shows the NBA team win/loss records at the close of the regular 2014–2015 season. Eight teams from each conference make the playoffs. For the season being studied, Western Conference teams were at a disadvantage because there were many strong teams in that conference. NBA schedules were not balanced.

Traditional unidimensional scaling converts multiple paired comparisons or rank orders to interval measures (Rounds et al., 1978) and represents an effective method for assessing the relative performance of teams with disparate schedules. We construct a matrix with the number of times each team beats each other team. This matrix is con-

3.5. RATING TEAMS WITH DISPARATE SCHEDULES

Table 3.2: NBA Team Records (2014–2015 Season)

Conference	Division	Team Name (Playoff *)	Abbreviation	Wins	Losses	P(Win)
Eastern	Atlantic	Boston Celtics*	BOS	40	42	0.488
		Brooklyn Nets*	BKN	38	44	0.463
		New York Knicks	NYK	17	65	0.207
		Philadelphia 76ers	PHI	18	64	0.220
		Toronto Raptors*	TOR	49	33	0.598
	Central	Chicago Bulls*	CHI	50	32	0.610
		Cleveland Cavaliers*	CLE	53	29	0.646
		Detroit Pistons	DET	32	50	0.390
		Indiana Pacers	IND	38	44	0.463
		Milwaukee Bucks*	MIL	41	41	0.500
	Southeast	Atlanta Hawks*	ATL	60	22	0.732
		Charlotte Hornets	CHA	33	49	0.402
		Miami Heat	MIA	37	45	0.451
		Orlando Magic	ORL	25	57	0.305
		Washington Wizards*	WAS	46	36	0.561
Western	Southwest	Dallas Mavericks*	DAL	50	32	0.610
		Houston Rockets*	HOU	56	26	0.683
		Memphis Grizzlies*	MEM	55	27	0.671
		New Orleans Pelicans*	NOP	45	37	0.549
		San Antonio Spurs*	SAS	55	27	0.671
	Northwest	Denver Nuggets	DEN	30	52	0.366
		Minnesota Timberwolves	MIN	16	66	0.195
		Oklahoma City Thunder	OKC	45	37	0.549
		Portland Trail Blazers*	POR	51	31	0.622
		Utah Jazz	UTA	38	44	0.463
	Pacific	Golden State Warriors*	GSW	67	15	0.817
		Los Angeles Clippers*	LAC	56	26	0.683
		Los Angeles Lakers	LAL	21	61	0.256
		Phoenix Suns	PHX	39	43	0.476
		Sacramento Kings	SAC	29	53	0.354

Source: Adapted from Miller (2016).

verted to a matrix of proportions, which is then averaged across rows or columns and referred to a standard normal distribution. As a final step in the process, we set a desired mean and standard deviation for scale scores. The method automatically adjusts for strength of schedule while defining interval-level measures from what were originally paired comparisons.

Figure 3.1 shows the results of a unidimensional scaling of NBA team records from the 2014–2015 season. A team of average strength is arbitrarily set to have a score of 500 on this scale, with the standard deviation across teams set to 100. Unidimensional scales reflect strength differences between teams, not just rank orders of teams.

Note the strength of the Golden State Warriors relative to other teams in Figure 3.1. From unidimensional scaling, we would have predicted the Golden State Warriors to be victorious in the playoffs for the 2014–2015 season, which they were.

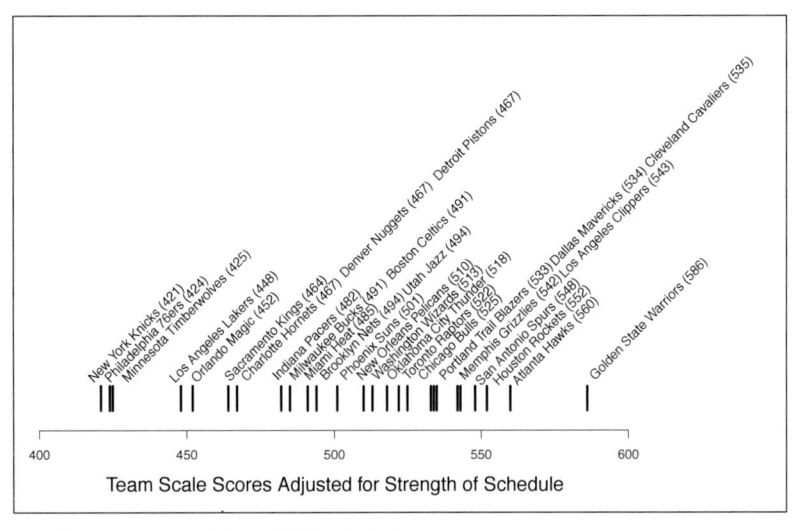

Source: Adapted from Miller (2016).

Figure 3.1: Assessing NBA Team Strength (2014–2015)

Unidimensional scaling builds on the method of paired comparisons, which has been used in taste-testing experiments, psychophysical investigations, preference scaling, and studies of inter-rater reliability, as well as in ranking players and teams. The extensive literature in this field, documented in Davidson and Farquhar (1976), dates back to early work in psychometrics by Thurstone (1927) and Guilford (1936) and in statistics by Bradley and Terry (1952). Miller (2008) showed how unidimensional scaling and win/loss metrics can be used to pick winning baseball teams.

3.6 Entropy Describes Repertoire

Returning to baseball, consider what is possible with counts of pitch types. All other things being equal, the more uncertain a batter is about the next pitch to be thrown, the less likely he is to have a productive at-bat. Batter uncertainty is a function of the number of pitches in a pitcher's arsenal and the extent to which the pitcher throws each pitch. Facing a rookie with a 100-mile-per-hour fastball, but only a fastball, may present a challenge to a batter. But add a change-up or curve ball to the mix, and that rookie pitcher presents a formidable challenge.

The catcher on the fielding team uses hand gestures, signalling to the pitcher which pitch to throw next. And, recognising the advantage of knowing which pitch will be thrown next, hitting teams often try to read the catcher's signs—a practice known as sign stealing, which is possible when the hitting team has a runner on second base. Successful sign stealing requires the runner on second to send his own sign to the batter about which pitch will be thrown next. Technology-assisted sign stealing is forbidden.

Pitch type is nominal, while the frequency distribution of pitches is a set of counts or ratio measures. To describe a pitcher's distribution of pitches, we can use information entropy, defined by

$$Entropy = -\sum_i P_i \, log_2 P_i$$

where P_i is the proportion of pitches of type i.

Entropy is a ratio measure showing the degree of uncertainty (information bits) associated with a pitcher's repertoire. Figure 3.2 provides examples of information entropy for pitchers with disparate pitch distributions. Notice how entropy increases as the number of pitches increases and as the frequency of pitches across pitch types is more evenly distributed.

Throwing only one pitch is aligned with the entropy zero point. Throwing two pitches equally often has entropy 1. And throwing four pitches equally often has entropy 2. While not a performance measure itself, entropy summarises a pitcher's repertoire and may well be related to pitching success.

Martin (2018) identified pitch types from pitch tracking data, the speed and location of pitches. In his initial explorations, Martin failed to find a strong relationship between entropy and pitching success, as measured by strikeout rate.

Entropy as a descriptor of player or team repertoire may be applied to various competitive sports. A team's playbook, with offensive

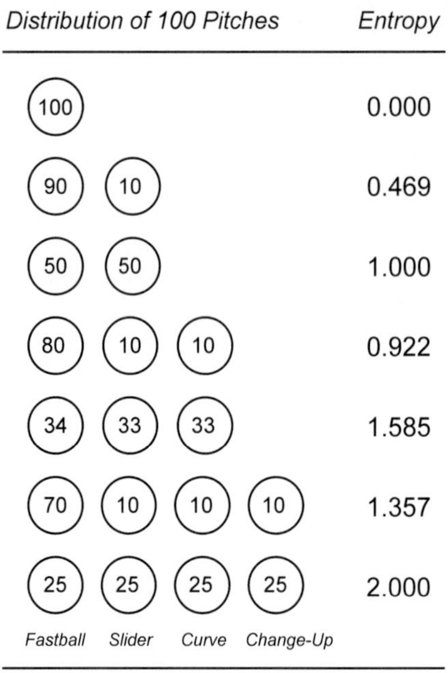

Distribution of 100 Pitches	Entropy
(100)	0.000
(90) (10)	0.469
(50) (50)	1.000
(80) (10) (10)	0.922
(34) (33) (33)	1.585
(70) (10) (10) (10)	1.357
(25) (25) (25) (25)	2.000
Fastball Slider Curve Change-Up	

Figure 3.2: Pitch Distributions and Information Entropy

and defensive alignments, represents its repertoire. More plays and more equally distributed play calling translate into higher entropy or uncertainty. We look to subsequent research to examine potential relationships between entropy and player or team performance.

3.7 The Embeddings Approach

Much of baseball strategy deals with pitcher-batter matchups, but what we know about specific pitcher-batter matchups is limited. Common pitcher performance measures are computed across all batters faced; just as common batter performance measures are computed across all pitchers faced. These aggregate measures say nothing about context or in-game situations involving a specific pitcher facing a specific batter.

We can disaggregate pitcher-batter data, drilling down or filtering by pitcher and batter. To advise managers about particular game situations or contexts, we can compute summary statistics on filtered

3.7. THE EMBEDDINGS APPROACH

data. For example, we could select all cases of a specific left-handed pitcher facing right-handed batters. But if we drill down to a specific pitcher and batter, we will be left with very small sample sizes. In fact, there are many pitcher-batter matchups with zero observations, such as matchups involving pitchers and batters from different leagues.

Major League Baseball Advanced Media has systems to track individual pitches, including their speed and trajectories, providing more detailed data on the tendencies of individual pitchers. In the aggregate the speed and location of a particular pitcher's pitches could be matched up against the types of pitches that a particular batter hits or misses. This sort of aggregate data research may hold promise in the future.

Working with baseball data, Alcorn (2018) and Miller (2018) show how to use neural networks to generate vector representations of pitchers and batters, providing more complete performance measures in context. These vector representations can then be used to evaluate teams and players, predict runs scored, and guide in-game strategy.

A neural network is a mathematical model composed of nodes and links. It represents a flexible, non-linear, data-adaptive model, a mathematical transformation or mapping from input nodes to output nodes. In a supervised learning context, input nodes represent explanatory variables and output nodes represent response variables we are trying to predict.

We try to match the structure of the neural network to the modelling task at hand. Moving from input nodes to internal (hidden) nodes and from internal nodes to output nodes, there are many network structures and mathematical transformations (activity functions) to consider. Dense or fully connected networks, for example, have every node in one layer connecting to every node in the next layer. A simple neural network may have only one internal layer of nodes, whereas a deep neural network (deep learning model) will have more than one internal layer (Géron, 2019; Goodfellow et al., 2016; Haykin, 1999).

In modelling pitcher-batter matchups, it makes sense to have as many input nodes as there are distinct pitchers and batters in the training data. For each pitcher all input node values are 0 except for one, which has the value 1. Likewise, for each batter, all input node values are 0 except for one, which has the value 1. This technique, known as one-hot encoding, provides a nominal representation of players. The only thing that we assume about players is that some are pitchers and others are batters.

Every link in a neural network is associated with a weight or parameter value that must be estimated. The objective of estimation is to minimise loss or errors in prediction. And what are we trying to predict? The outcome of each play. We use observed events from

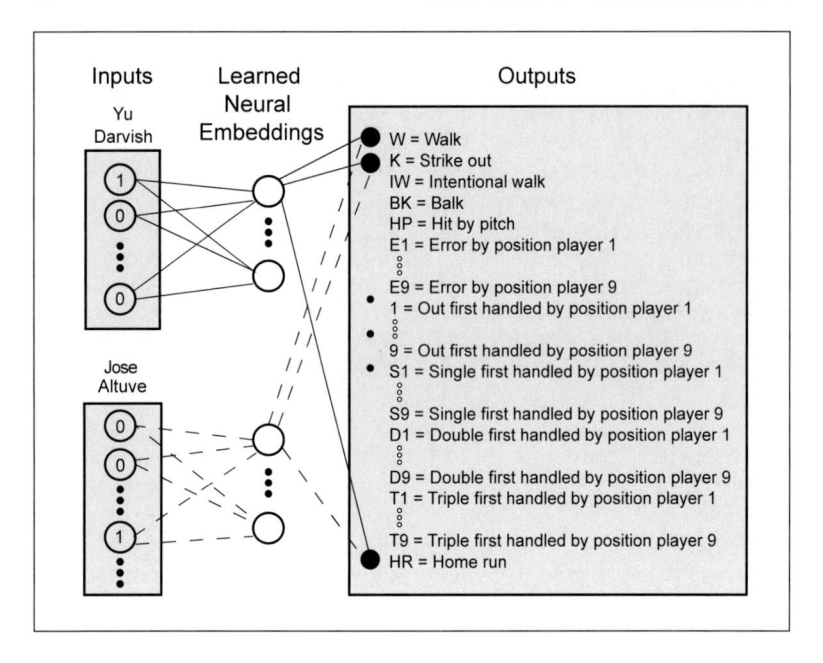

Figure 3.3: Neural Network Structure for Player Embeddings

past games to represent a multinomial outcome—these are the output node values in the training data. Multinomial prediction utilises softmax, a special mathematical function between internal nodes and output nodes. Softmax provides an estimation of the probability of each outcome.

A simple neural network structure is employed in the study reviewed here. A single internal layer is defined with separate pitcher and batter nodes. Fitted weights on these nodes represent separate embeddings for pitchers and batters, respectively.

Neural network embeddings comprise a radical shift in the way we think about performance measurement in baseball. The embeddings model requires no player metrics in predicting play-by-play outcomes. The inputs to the model are minimal—we know only player names (via one-hot encodings) and the fact that some players are pitchers and others batters.

Figure 3.3 from Miller (2018) summarises the neural network structure used to obtain pitcher and batter embeddings. The example shows a pitcher-batter matchup from the 2017 World Series between the Los Angeles Dodgers and Houston Astros.

3.7. THE EMBEDDINGS APPROACH

Table 3.3: Embeddings for Dodger Starting Pitchers

Player	W1	W2	W3	W4	W5	W6	W7	W8	W9
Yu Darvish	0.518	0.532	0.451	0.459	0.563	0.460	0.518	0.509	0.570
Rich Hill	0.491	0.550	0.526	0.492	0.538	0.533	0.639	0.453	0.528
Clayton Kershaw	0.504	0.566	0.444	0.628	0.534	0.522	0.521	0.433	0.615
Kenta Maeda	0.480	0.518	0.510	0.479	0.541	0.457	0.452	0.463	0.510
Alex Wood	0.511	0.475	0.508	0.575	0.532	0.530	0.589	0.507	0.494

Table 3.4: Embeddings for the Astros Position Players

Player	W10	W11	W12	W13	W14	W15	W16	W17	W18
Jose Altuve	0.527	0.365	0.534	0.458	0.471	0.312	0.487	0.408	0.373
Alex Bregman	0.602	0.428	0.499	0.508	0.604	0.443	0.443	0.527	0.462
Carlos Correa	0.454	0.430	0.463	0.518	0.569	0.438	0.522	0.345	0.530
Marwin Gonzalez	0.429	0.464	0.528	0.499	0.451	0.524	0.495	0.493	0.480
Yulieski Gurriel	0.516	0.425	0.614	0.445	0.580	0.374	0.500	0.538	0.433
Brian McCann	0.467	0.441	0.564	0.513	0.494	0.551	0.417	0.512	0.644
Josh Reddick	0.508	0.485	0.605	0.490	0.478	0.571	0.412	0.501	0.544
George Springer	0.531	0.398	0.391	0.486	0.557	0.408	0.554	0.391	0.506

The structure of the neural network employed by both Alcorn (2018) and Miller (2018) was set with only nine interior nodes for the pitcher and nine interior nodes for the batter. Accordingly, the fitted embeddings represented numerical vectors with nine values for each pitcher and nine values for each batter.

Miller (2018) used Retrosheet (2018) plate appearance data for all regular season MLB games from 2013–2017 to train the neural network model. Each observation or plate appearance constituted a pitcher-batter matchup—the context of play. The result of each plate appearance was a multinomial outcome with one of 52 possible values. (Note that 51 distinct output values were observed in the study reviewed here (Miller, 2018).) A softmax loss function was defined, as appropriate for a multinomial response. Pitcher and batter embeddings were obtained for all players with sufficient data, including starting pitchers and starting lineup position players from the 2017 World Series.

Tables 3.3 and 3.4 show the estimated embeddings for five possible Los Angeles Dodger starting pitchers and a starting lineup of eight Houston Astro position players, respectively, as anticipated prior to the seventh game of the 2017 World Series.

By taking an embeddings approach to measurement, we go directly from events on the field, describing pitcher-batter matchups and outcomes, to measures that reflect the behaviour or performance of players. Each pitcher in this study is represented by nine numbers. Each batter is represented by nine numbers. We have no words to

Table 3.5: Embeddings-Driven Outcome Probabilities

Pitcher	Batter	Out Fielder 1	Out Fielder 2	● ● ●	Strikeout	Walk
Yu Darvish	Jose Altuve	0.0189	0.0070		0.1646	0.0690
Yu Darvish	Alex Bregman	0.0122	0.0060		0.2634	0.0989
Yu Darvish	Carlos Correa	0.0158	0.0055		0.2820	0.1204
Yu Darvish	Marwin Gonzalez	0.0242	0.0081		0.2925	0.0689
Yu Darvish	Yulieski Gurriel	0.0184	0.0072		0.1673	0.0532
Yu Darvish	George Springer	0.0119	0.0050		0.2528	0.0963
Yu Darvish	Brian McCann	0.0165	0.0057		0.2124	0.0984
Yu Darvish	Josh Reddick	0.0136	0.0056		0.3471	0.1154
Rich Hill	Jose Altuve	0.0232	0.0062		0.1499	0.0738
Rich Hill	Alex Bregman	0.0153	0.0054		0.2447	0.1080
Rich Hill	Carlos Correa	0.0198	0.0049		0.2636	0.1323
Rich Hill	Marwin Gonzalez	0.0313	0.0075		0.2808	0.0778
Rich Hill	Yulieski Gurriel	0.0229	0.0064		0.1547	0.0579
Rich Hill	George Springer	0.0156	0.0047		0.2460	0.1102
Rich Hill	Brian McCann	0.0215	0.0054		0.2058	0.1121
Rich Hill	Josh Reddick	0.0171	0.0051		0.3237	0.1265
Clayton Kershaw	Jose Altuve	0.0246	0.0062		0.1548	0.0395
Clayton Kershaw	Alex Bregman	0.0165	0.0055		0.2559	0.0585
Clayton Kershaw	Carlos Correa	0.0215	0.0051		0.2783	0.0724
Clayton Kershaw	Marwin Gonzalez	0.0330	0.0075		0.2881	0.0414
Clayton Kershaw	Yulieski Gurriel	0.0237	0.0063		0.1556	0.0302
Clayton Kershaw	George Springer	0.0169	0.0048		0.2599	0.0603
Clayton Kershaw	Brian McCann	0.0232	0.0055		0.2158	0.0609
Clayton Kershaw	Josh Reddick	0.0186	0.0052		0.3432	0.0695
Kenta Maeda	Jose Altuve	0.0225	0.0069		0.1162	0.0517
Kenta Maeda	Alex Bregman	0.0154	0.0062		0.1959	0.0781
Kenta Maeda	Carlos Correa	0.0201	0.0057		0.2131	0.0966
Kenta Maeda	Marwin Gonzalez	0.0304	0.0083		0.2176	0.0545
Kenta Maeda	Yulieski Gurriel	0.0218	0.0069		0.1174	0.0397
Kenta Maeda	George Springer	0.0149	0.0052		0.1882	0.0761
Kenta Maeda	Brian McCann	0.0203	0.0058		0.1550	0.0762
Kenta Maeda	Josh Reddick	0.0178	0.0060		0.2684	0.0947
Alex Wood	Jose Altuve	0.0272	0.0058		0.0997	0.0536
Alex Wood	Alex Bregman	0.0192	0.0054		0.1734	0.0835
Alex Wood	Carlos Correa	0.0249	0.0050		0.1873	0.1025
Alex Wood	Marwin Gonzalez	0.0387	0.0074		0.1969	0.0595
Alex Wood	Yulieski Gurriel	0.0266	0.0060		0.1019	0.0416
Alex Wood	George Springer	0.0194	0.0047		0.1737	0.0849
Alex Wood	Brian McCann	0.0263	0.0053		0.1425	0.0847
Alex Wood	Josh Reddick	0.0221	0.0052		0.2372	0.1011

describe pitching and batting performance—no words, just numbers.

With embeddings in hand we compute outcome probability vectors for each potential pitcher-batter matchup, as shown in Table 3.5 for the seventh game of the 2017 World Series. Fifty-one (51) distinct events were observed in the training data, beginning with an out first fielded by the player in position 1 (the pitcher) and ending with a walk. So there are 51 expected event probabilities for each pitcher-batter combination. These event probabilities may then be used to drive game-day simulations.

Consider how Dave Roberts, manager for the Los Angeles Dodgers for the 2017 season, might have used embeddings in selecting a starting pitcher for the seventh game of the World Series. Embeddings yield expected event probabilities for each pitcher-batter matchup, and these expected event probabilities can be used to drive a Monte Carlo simulation.

3.7. THE EMBEDDINGS APPROACH

Table 3.6: Who should pitch for the Dodgers in game seven?

Pitcher	Days Rested	Series ERA	Predicted ERA
Clayton Kershaw	2	5.40	3.10
Yu Darvish	4	21.60	4.02
Rich Hill	0	2.08	4.65
Alex Wood	3	3.48	4.73
Kenta Maeda	0	1.59	4.74

In running simulated games for each Dodger pitcher, we make a few simplifying assumptions, such as Astros pitchers make an out when batting, there are no stolen bases and no sacrifices, and all errors are one-base errors. An additional assumption is that on-base runners advance no further than the batter. That is, when an Astros batter hits a single, all on-base runners advance one base. When a batter hits a double, all on-base runners advance two bases. And when a batter hits a triple, all on-base runners advance three bases.

We execute two thousand simulated games under these assumptions, obtaining starting pitcher performance estimates displayed in Table 3.6, along with each pitcher's number of days rest and series earned run average (ERA) just prior to the seventh game. Earned run average, a summary measure of pitching prowess, represents the expected number of runs scored against a pitcher across nine innings, assuming that no errors have been made by the fielding team.

What actually happened in the seventh game of the 2017 World Series? Dave Roberts selected Yu Darvish as the starting pitcher. Darvish gave up 5 runs (4 of them earned) while pitching only one and two-thirds of an inning. Relief pitchers Brandon Morrow, Clayton Kershaw, Kenley Jansen, and Alex Wood gave up no runs through the rest of the game, a game the Dodgers lost 5-to-1. Most notable among the Dodgers pitchers was Clayton Kershaw, who completed a no-run, four-inning stretch across innings 3 through 6. Dave Roberts' starting pitcher selection was not guided by an analysis of neural network embeddings, but it could have been—and for Dodgers fans, maybe it should have been.

Managerial decisions can benefit from an embeddings approach to sports performance measurement. A key advantage of embeddings, as shown in this example, is its systematic approach to predicting what is likely to happen with selected pitcher-batter matchups, including those for which there are limited or no historical data, which is often the case when National League pitchers face American League batters.

Sports analytics should be more than a matter of manipulating box scores and play-by-play statistics. With detailed on-field or on-court data from every sporting contest, teams face challenges in data management, data engineering, and analytics. Data-driven, data-adaptive measurement methods such as neural network embeddings offer a path forward, a way to utilise the growing data around sports.

3.8 Validity—Words that Matter

The words we use for some measures are little more than descriptions of measurement methods. The meaning of these measures is clear. Other measures require more explanation. Heart rate and blood pressure, for example, are easily understood. Resilience, strength, and endurance are more difficult to define, and measure—the meaning of such measures may be uncertain or subject to debate.

Returning to baseball, most analysts would agree that batting average (BA) and on-base percentage (OBP) are easy to explain. They are computed as simple proportions drawing on in-game events. Clutch hitting or pitching efficiency, on the other hand, would be more difficult to compute and explain.

When we talk about *validity*, we are thinking of the degree to which a measure measures what it is said to measure. There are subjective assessments of *face validity* or *content validity*. We examine measurements to see the degree to which they appear to measure the attributes they are said to measure. Taking a more objective approach to validity, we try to demonstrate *predictive validity*. Knowing how two attributes or traits are related in theory, we can create measures of those traits and examine the degree to which their corresponding measures relate as theory suggests. The meaning of a measure is defined by its relationship to other measures—this is *construct validity*, the logical extension of predictive validity.

Construct validity is illustrated by the multitrait-multimethod matrix and what Campbell and Fiske (1959) call convergent validity and discriminant validity. Convergent validity refers to the idea that different measures of the same trait should converge. That is, different measures of the same trait or attribute should have relatively high correlations. Discriminant validity refers to the notion that measures of different traits should diverge. In other words, measures of different traits should have lower correlations than measures of the same trait.

A prototypical measurement study following Cambell and Fiske involves a multitrait-multimethod matrix, a matrix with rows and columns associated with traits (attributes) and methods (measurement procedures). Each element of the matrix represents a trait-method unit. Figure 3.4 shows a hypothetical multitrait-multimethod matrix with four baseball measures.

What do we expect to see in the multitrait-multimethod matrix

3.8. VALIDITY—WORDS THAT MATTER

The meaning of a measure is defined by its relationships to other measures.

		Method 1: Practice Measures		Method 2: Game-Day Measures	
		Trait A: Hitting Ability	Trait B: Power Hitting Ability	Trait A: Hitting Ability	Trait B: Power Hitting Ability
Method 1: Practice Measures	Trait A: Hitting Ability	0.90			
	Trait B: Power Hitting Ability	0.40	0.90		
Method 2: Game-Day Measures	Trait A: Hitting Ability	0.45	0.20	0.80	
	Trait B: Power Hitting Ability	0.20	0.50	0.30	0.80

Convergent validity is demonstrated by high correlations in the validity diagonal.
Discriminant validity is demonstrated by the relative sizes of correlations in diagonals and triangles:

Reliability diagonal > Validity diagonal > Heterotrait-monomethod triangle > Heterotrait-heteromethod triangle

Method variance is demonstrated by relatively high heterotrait-monomethod correlations when traits are assumed to be uncorrelated.

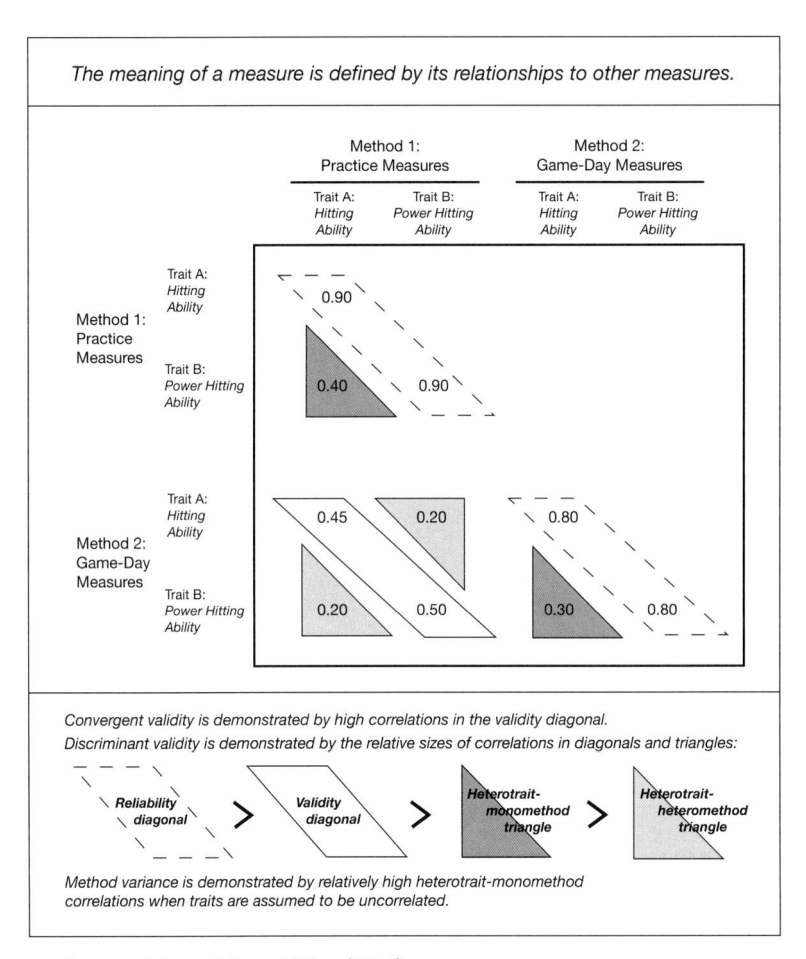

Source: Adapted from Miller (2016).

Figure 3.4: Multitrait-Multimethod Matrix for Baseball Measures

Table 3.7: A Measurement Model for Sports

Physiological	Physical	Psychological	Behavioral	Environmental
Blood Pressure	Agility	Anxiety	Nutrition	Built Environment
Glucose and Insulin	Anaerobic Power	Competitiveness	Sleep	Social Support Groups (Coaches, Parents, Peers)
Heart Rate Variability	Balance	Confidence	Substance Use	Socioeconomic Status
Lactate Threshold	Body Composition	Depression		
Methylome	Cardiorespiratory Endurance	Impulsiveness		
Previous Injuries	Coordination Ability	Intellect of Sport		
Respiratory Rate	Flexibility	Motivation		
Resting Heart Rate	Muscular Endurance	Narcissism		
Telomere Length	Muscular Power	Perfectionism		
Vision	Muscular Strength	Resiliency		
VO_2 max	Reaction Time	Self-Efficacy		
	Sport-Specific Skills	Self-Esteem		
		Vigor		

Source: Adapted from Martin (2016).

in Figure 3.4? The reliability diagonals show measures of the same traits by the same methods. Assessed through test-retest reliability or internal consistency measures, values on the reliability diagonal represent nearly perfect correlations.

We should see different measures of the same trait correlating positively on the validity diagonal. Hitting in batting practice should have a positive correlation with hitting in games. We expect measures of the same trait to correlate more highly with one another than with measures of different traits. Accordingly, we should see higher correlations on the validity diagonal than in either the heterotrait-monomethod or the heterotrait-heteromethod triangles.

Discussions of validity touch on fundamental issues in philosophy of science—issues of theory construction, measurement, and testability. There are no easy answers here. If the theory is correct and the measures valid, then the pattern of relationships among the measures should be similar to the pattern predicted by theory. To the extent that this is true for observed data, we have partial confirmation of the theory and, at the same time, demonstration of construct validity.

If predictions from theory do not pan out using selected measures, then we are faced with a dilemma. The theory could be wrong, one or more of the measures could be invalid, or we could have observed an event of low probability with correct theory and valid measures.

In developing predictive models of performance in sport, there are many factors to consider. A comprehensive measurement regimen would include physiological, physical, psychological, behavioural, and environmental variables, as shown in Table 3.7.

We acknowledge the importance of words and numbers in science. With a traditional approach to measurement, we begin with words describing player attributes. We assign numbers to player attributes, and we make predictions, relying on models that relate player attributes to game outcomes.

As the embeddings approach demonstrates, however, we can make predictions without defining player attributes, without theories of how attributes relate to game outcomes. To get embeddings, we merely note who has played in past games and observe the events of those games. Words, to the extent that we use words at all, flow from the interpretation of embeddings.

3.9 Conclusions

Regarding measurement and philosophy of science, consider the umpire story, developed long before replay technologies were available:

> After a long day of disputed calls at the ballpark, three umpires are asked to justify their methods. The first umpire, an empiricist by persuasion, says, *I call them as I see them*. The second, with the faith of a philosophical realist, replies, *I call them as they are*. Not to be outdone, the third umpire, with the self-proclaimed authority of an operationist or logical empiricist, says, *The way I call them—that's the way they are*.

We are the umpires in science meets sports. We decide on the measurement approach and on a rationale for justifying that approach. A fundamental objective of science is prediction—showing that one set of measures (explanatory variables or inputs) relates to another set of measures (response variables or outputs). We can realise that objective using either a traditional or embeddings approach to measurement.

Traditional measurement, as commonly defined, deals with the assignment of numbers to attributes according to rules. We show how player attributes, described with words, and measured with numbers, relate to athletic performance and game outcomes.

With the embeddings approach to measurement, we assign numbers to players directly. The only attribute to speak of is the player's role or position in the game. The mantra of the embeddings approach may be simply stated: "Numbers first. Words can come later."

Before we throw away a hundred years of measurement theory, however, it would be wise to acknowledge that words facilitate thinking. Whether we take a traditional or embeddings approach, words help us to understand the meaning of scientific research.

Bibliography

Albert, J. H. (2009). *Bayesian Computation with R*. Springer, New York.

Albert, J. H. and Bennett, J. (2001). *Curve Ball: Baseball, Statistics, and the Role of Chance in the Game*. Springer, New York.

Alcorn, M. A. (2018). *[Batter-Pitcher]2Vec: Statistic-Free Talent Modeling with Neural Player Embeddings*. MIT Sloan Sports Analytics Conference, Boston. Retrieved from the World Wide Web on September 30, 2019 at https://github.com/airalcorn2/batter-pitcher-2vec.

Appleton, D. R. (1995). May the best man win? *The Statistician*, 44(4):529–538.

Baker, B. O., Hardyck, C. D., and Petrinovich, L. F. (1966). Weak measurements vs. strong statistics: An empirical critique of S. S. Stevens' proscriptions on statistics. *Educational and Psychological Measurement*, 26:291–309.

Baumer, B. and Zimbalist, A. (2014). *The Sabermetric Revolution: Assessing the Growth of Analytics in Baseball*. University of Pennsylvania Press, Philadelphia.

Baumer, B. S., Jensen, S. T., and Matthews, G. J. (2015). OpenWAR: An open source system for evaluating overall player performance in Major League Baseball. *Journal of Quantitative Analysis in Sports*, 11(2):69–84.

Bradley, R. A. and Terry, M. E. (1952). Rank analysis of incomplete block designs: I. The method of paired comparisons. *Biometrika*, 39(3/4):324–345.

Campbell, D. T. and Fiske, D. W. (1959). Convergent validity and discriminant validity by the multitrait-multimethod matrix. *Psychological Bulletin*, 56:81–105.

Carlin, B. P. (1996). Improved NCAA basketball tournament modeling via point spread and team strength information. *The American Statistician*, 50(1):39–43.

Davidson, R. R. and Farquhar, P. H. (1976). A bibliography on the method of paired comparisons. *Biometrics*, 32(2):241–252.

Englemann, J. (2017). Possession-based player performance analysis in basketball (adjusted $+/-$ and related concepts). In Albert, J. H., Glickman, M. E., Swartz, T. B., and Koning, R. H., editors, *Handbook of Statistical Methods and Analyses in Sports*, pages 215–227. CRC Press, Boca Raton, Florida.

Fair, R. C. (2008). Estimated age effects in baseball. *Journal of Quantitative Analysis in Sports*, 4(1):1–39.

Géron, A. (2019). *Hands-On Machine Learning with Scikit-Learn, Keras, and TensorFlow: Concepts, Tools, and Techniques to Build Intelligent Systems*. O'Reilly, Sebastopol, California, second edition.

Goodfellow, I., Bengio, Y., and Courville, A. (2016). *Deep Learning*. MIT Press, Cambridge, Massachusetts.

Groeneveld, R. A. (1990). Ranking teams in a league with two divisions of t teams. *The American Statistician*, 44(4):277–281.

Guilford, J. P. (1936). *Psychometric Methods*. McGraw-Hill, New York.

Haykin, S. (1999). *Neural Networks: A Comprehensive Foundation*. Prentice-Hall, Upper Saddle River, New Jersey, second edition.

Langville, A. N. and Meyer, C. D. (2012). *Who's #1: The Science of Rating and Ranking*. Princeton University Press, Princeton, New Jersey.

Lewis, M. (2003). *Moneyball: The Art of Winning an Unfair Game*. W. W. Norton & Company, New York.

Ley, C., Van de Wiele, T., and Van Eetvelde, H. (2019). Ranking soccer teams on basis of their current strength: A comparison of maximum likelihood approaches. *Statistical Modelling*, 19(1):55–73.

Marchi, M. and Albert, J. H. (2014). *Analyzing Baseball Data with R*. CRC Press, Boca Raton, Flordia.

Martin, E. P. (2018). *Predicting Major League Baseball Strikeout Rates from Differences in Velocity and Movement Among Player Pitch Types*. Master's Thesis, Northwestern University, Evanston, Ill.

Martin, L. (2016). *Sports Performance Measurement and Analytics: The Science of Assessing Performance, Predicting Future Outcomes, Interpreting Statistical Models, and Evaluating the Market Value of Athletes*. Pearson Education, Old Tappan, New Jersey.

Miller, T. W. (2008). *Without a Tout: How to Pick a Winning Team*. Research Publishers, Manhattan Beach, California.

Miller, T. W. (2016). *Sports Analytics and Data Science: Winning the Game with Methods and Models*. Pearson Education, Old Tappan, New Jersey.

Miller, T. W. (2018). *Working with the Data of Sports.* Strata Data Conference, San Jose, California. Abstract and presentation slides at https://conferences.oreilly.com/strata/strata-ca-2018/public/schedule/detail/63339.

Oliver, D. (2004). *Basketball on Paper: Rules and Tools for Performance Analysis.* Potomac Books, Washington, D.C.

Retrosheet (2018). Retrosheet event files. Retrieved from the World Wide Web on March 4, 2018 at https://www.retrosheet.org/game.htm.

Rounds, J. B., Miller, T. W., and Dawis, R. V. (1978). Comparability of multiple rank order and paired comparison methods. *Applied Psychological Measurement,* 2(3):415–422.

Silver, N. (2004). Baseball prospectus basics: The science of forecasting. *Baseball Prospectus.* Retrieved from the World Wide Web, September 27, 2015, at http://www.baseballprospectus.com/article.php?articleid=2659/.

Silver, N. (2012). *The Signal and the Noise: Why So Many Predictions Fail—But Some Don't.* The Penguin Press, New York.

Sports Reference LLC (2019a). 2014 Major League Baseball team statistics and standings: Team & league standard batting. Retrieved from the World Wide Web on September 29, 2019 at https://www.baseball-reference.com/leagues/MLB/2014-standard-batting.shtml.

Sports Reference LLC (2019b). 2019 Major League Baseball team statistics and standings: Team & league standard batting. Retrieved from the World Wide Web on September 29, 2019 at https://www.baseball-reference.com/leagues/MLB/2019-standard-batting.shtml.

Stevens, S. S. (1946). On the theory of scales of measurement. *Science,* 103(2684):677–680.

Tango, T. M., Lichtman, M. G., and Dolphin, A. E. (2007). *The Book: Playing the Percentages in Baseball.* TMA Press.

Thompson, M. (1975). On any given Sunday: Fair competitor orderings with maximum likelihood methods. *Journal of the American Statistical Association,* 70(351):536–541.

Thurstone, L. L. (1927). A law of comparative judgment. *Psychological Review,* 34:273–286.

Torgerson, W. S. (1958). *Theory and Methods of Scaling.* Wiley, New York.

Van Eetvelde, H. and Ley, C. (2019). Ranking methods in soccer. In Kenett, R. S., Longford, T. N., Piegorsch, W., and Ruggeri, F., editors, *Wiley StatsRef: Statistics Reference Online*. Wiley, Hoboken, New Jersey.

West, B. T. (2006). A simple and flexible rating method for predicting success in the NCAA basketball tournament. *Journal of Quantitative Analysis in Sports*, 2(3):1–14.

Wikipedia (2019). Plus-minus. Retrieved from the World Wide Web on November 30, 2019 at https://en.wikipedia.org/wiki/Plus%E2%80%93minus.

Winston, W. L. (2009). *Mathletics: How Gamblers, Managers, and Sports Enthusiasts Use Mathematics in Baseball, Basketball, and Football*. Princeton University Press, Princeton, New Jersey.

Chapter 4
Analysing Positional Data

ULF BREFELD
LEUPHANA UNIVERSITY OF LÜNEBURG

JAN LASEK
INSTITUTE OF COMPUTER SCIENCE, POLISH ACADEMY OF SCIENCES

SEBASTIAN MAIR
LEUPHANA UNIVERSITY OF LÜNEBURG

Abstract

This chapter contains an introduction to processing positional data using football as an example. We present mathematical and computational models of increasing complexity to finally devise intelligent retrieval to assist video analysts. On the path to this goal, we discuss the computation of heat maps and the extraction of velocities and acceleration from (sequences of) positions. A combination of position and speed over time then allows to devise data-driven movement models, which, in turn, are a precondition to compute zones of control. The zones are then used to compute the relevancy of a situation for a given query in retrieval tasks.

4.1 Introduction

Recent advances in computer vision and machine learning allow not only to record data in sports at large scales but also to make sense of that data or turn it into prediction models. As a consequence, data are being collected in many different sports, including football, basketball, hockey, American football, or ski racing (see, e.g., Bornn et al. 2018; Harmon et al. 2016; Zhao et al. 2016), either by computer vision systems using (multiple) cameras (Manafifard et al., 2017) or by sensors attached to the athletes (Haase and Brefeld, 2014; Pettersen et al., 2014). Data captured regularly in football matches, such as the German Bundesliga, view the pitch as a coordinate system and comprise two-dimensional x/y positions of players, referee, and ball, recorded at either 10 or 25 frames per second, depending on the provider.[1] The data are usually accompanied by an event log containing basic events such as passes, corner kicks, shots, etc. While the latter can be parsed with standard techniques, positional data have their own spatio-temporal structure and processing positional data are somewhat more delicate.

In the course of this chapter, we show how to make sense of positional data and how to use that complex data to compute quantities of interest. Focusing on a player, we begin with sequences of her/his positions (trajectories) to compute heat maps showing densities of her/his whereabouts. An additional extraction of velocities and acceleration allows to devise data-driven movement models (Brefeld et al., 2019). Applying the latter to a situation at hand results in zones of control (Taki and Hasegawa, 2000) which can be analysed and processed further. As an exemplary application, we discuss their value for intelligent retrieval at the end of this chapter. The next section briefly reviews related work dealing with player positions and controlled zones.

4.2 Related Work

Player motion or movement models are important quantities for advanced match analyses. The simplest version of a movement model is given by the outer-graph of player positions, so that every position on the pitch is assigned to the closest player. The result is also known as a Voronoi tessellation with players as centres (Fonseca et al.,

---

[1] Some providers also support z-coordinates or even 3D-shapes of the athletes, however, in this chapter we will focus on a two-dimensional representation as it is the most general representation.

2012; Voronoi, 1908). Though their frequent deployment, Voronoi tessellations thus implement the assumption that every player can run equally fast in any direction, independent of her/his actual velocity and direction of movement. It is actually a very crude model of reality and often leads to implausible results.

More sophisticated approaches quantify the probability of attaining a certain position or compute the required amount of time to reach it. Traditional movement models are often based on simplified laws of physics (Fujimura and Sugihara, 2005; Taki and Hasegawa, 2000; Taki et al., 1996). Recently, Gudmundsson and Wolle (2014) have pondered about a movement model that is generated from observations only and Brefeld et al. (2019) propose the first purely data-driven movement model. One of the advantages lies in the personalisation. In contrast to one-serves-all models derived by simplifying laws of physics, data-driven models can be trivially computed on only data from a single player, therefore, capturing her/his individual traits. A hybrid approach is proposed by Spearman et al. (2017), where a physical movement model is calibrated using positional data.

4.3 Player Movement

4.3.1 Positions

Positional data of the players and the ball are typically described by a sequence of positions $\mathcal{T} = (\mathbf{p}_t)_{t \in T}$ in two-dimensional space. Here, $\mathbf{p}_t = (x_t, y_t) \in \mathbb{R}^2$ denotes a position at time t within a set T of timestamps. The timestamps are usually equidistant and stem from the frames per second of the data recording technique at hand. A sequence of positions is sometimes called a *trajectory*.

A game of 90 minutes contains over $135,000$ positions for every player. Unfortunately, the sheer quantity renders straightforward visualisations of the trajectories as in the left-hand side of Figure 4.1 uninterpretable. However, the raw input can be easily turned into a heat map to highlight densities of whereabouts of a player. To compute a heat map, the pitch is usually discretised into equidistant patches. For every patch the number of positions lying in the respective area are counted. The right-hand side of Figure 4.1 depicts the corresponding heat map. Such visualisations of player movements are increasingly popular tools during broadcasts and for post-match player performance analysis.

4.3.2 Velocities and Accelerations

Though the heat map in Figure 4.1 (right) visualises whereabouts of a player, an important quantity is ignored: time. Imagine two players moving on a line: One of them slowly moves from one end to the other while the other player uses the same time to run back and forth as often as she/he can. The two corresponding heat maps are identical as both players realise a uniform density on the line, but their movements are actually quite different.

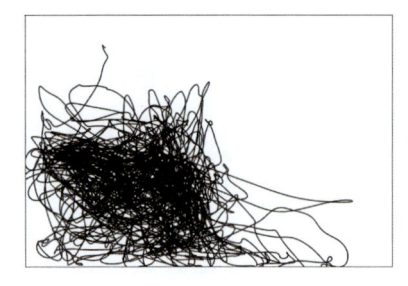

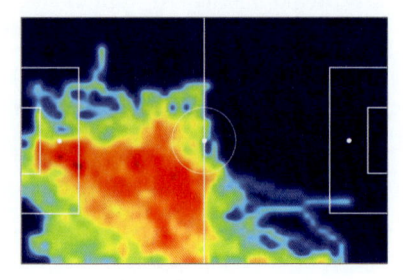

Figure 4.1: Trajectory (left) and heat map (right) of a player.

We now focus on the extraction of velocities and acceleration. Given a sequence of positions $\mathcal{T}$, the velocity of the player at time t is given by the derivative of the position with respect to time and can be estimated by
$$\mathbf{v}_t = (\mathbf{p}_{t+t_\Delta} - \mathbf{p}_{t-t_\Delta})/(2t_\Delta)$$
for some sufficiently small t_Δ. Hence, the scalar velocity or speed is $v_t = \|\mathbf{v}_t\|_2$. Assuming that the positions are represented in meters, the unit of the velocity is m/s and can be converted into km/h by a multiplication with 3.6.

Throughout a match, players on different positions exhibit different velocity patterns. Figure 4.2 shows an example for a goalkeeper, a defender, and a striker. In the figure we defined standing as $< 1\ km/h$, walking as $1-7\ km/h$, jogging as $7-14\ km/h$, running as $14-20\ km/h$ and sprinting as $> 20\ km/h$.[2] While infield players have a similar distribution of speed values, the goalkeeper barely runs during a match.

Another characteristic trait quantifying a player's ability to move is her/his acceleration in a given direction. The acceleration vector is the derivative of the velocity vector with respect to time. Hence, it can be estimated from data similarly as the velocity vectors. Figure 4.3 visualises the acceleration profile for a player moving in the direction of the x-axis with different velocities. The figure shows that the higher the speed, the more difficult it is to accelerate further in the direction of the movement.

---

[2] In practice, the intervals may depend on the application at-hand.

4.3. PLAYER MOVEMENT

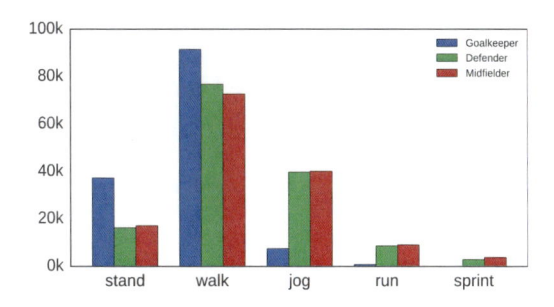

Figure 4.2: Distribution of velocities.

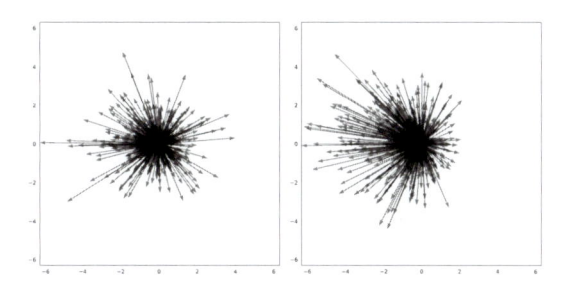

Figure 4.3: Exemplary acceleration vectors (m/s^2) in x/y-space for a player moving within $5-10\ km/h$ (left) and $14-25\ km/h$ (right).

4.3.3 Movement Models

Combining positions and velocities directly leads to movement models. A movement model lays the ground for sophisticated analyses and describes the ability of a player to move in a given time horizon $t_\Delta > 0$ relative to her/his current position. A movement model needs to be generally applicable and dependent on the actual speed or direction of movement. We thus need to record the player's velocity $\mathbf{v}_t$ at the time of application as described in Section 4.3.2.

Given the velocity, the correct movement model is retrieved and rotated such that it fits the direction of movement at time t. This is done by transforming every subsequent position within the trajectory into a local coordinate system, relative to the current position (Brefeld et al., 2019). Let $(\mathbf{p}_s, \mathbf{p}_t, \mathbf{p}_u)$ be a triplet of positions within a player's trajectory with $s < t < u$, $t - s = t_\delta$ and $u - t = t_\Delta$. The variable t_δ denotes the time span that is used for the estimation of the direction of movement of a player. In practice, it can be set to some small value, e.g., we used 0.2 seconds throughout this chapter, however, larger

---
Algorithm 1 Computation of movement samples.

Input: Data set $\mathcal{D} = \{(\mathbf{p}_{t_i}, v_{t_i})\}_{i=1}^{n}$, velocity range V
Output: Set $\mathcal{S}_{t_\Delta, V}$ of attained positions in time t_Δ
1: **for** $s < t < u$ satisfying $s = t - t_\delta, u = t + t_\Delta$ **do**
2:    **if** $v_t \in V$ **then**
3:       $\mathbf{p} = \psi(\mathbf{p}_s, \mathbf{p}_t, \mathbf{p}_u)$ ▷ transformed destination as in Eq. (4.1)
4:       $\mathcal{S}_{t_\Delta, V} = \mathcal{S}_{t_\Delta, V} \cup \{(\mathbf{p}, v_t)\}$ ▷ append the sample to $\mathcal{S}_{t_\Delta, V}$
5:    **end if**
6: **end for**
---

values may be supported by the application at hand. We use the coordinates $\mathbf{p}_s$ and $\mathbf{p}_t$ to estimate the direction in which the player is moving, while $\mathbf{p}_u$ is used to estimate her/his ability to move. First, the triplet is transformed by a translation such that $\mathbf{p}_t$ is centred at $(0,0)$. Then, we rotate the triplet such that the vector $\overrightarrow{\mathbf{p}_s\mathbf{p}_t} = (x_t - x_s, y_t - y_s)$ is aligned with the x-axis. This way, the transformed position $\mathbf{p}_u$ describes the point which the player reaches in time t_Δ, assuming her/his current position is the origin, moving in direction of x-axis with a given speed of $v_t = \|\mathbf{v}_t\|_2$.

We denote this transformation with a function

$$\psi : \mathbb{R}^2 \times \mathbb{R}^2 \times \mathbb{R}^2 \to \mathbb{R}^2, \quad (\mathbf{p}_s, \mathbf{p}_t, \mathbf{p}_u) \mapsto \mathbf{p} = (x, y) \tag{4.1}$$

and obtain the transformed point $\mathbf{p} = (x, y)$ using a representation in polar coordinates $(x, y) = (r \cdot \cos(\theta), r \cdot \sin(\theta))$, where θ is a signed angle and r is the distance. The angle θ can be computed as

$$\theta = \measuredangle(\overrightarrow{\mathbf{p}_s\mathbf{p}_t}, \overrightarrow{\mathbf{p}_t\mathbf{p}_u})$$
$$= \operatorname{atan2}(y_u - y_t, x_u - x_t) - \operatorname{atan2}(y_t - y_s, x_t - x_s)$$

for $\mathbf{p}_s \neq \mathbf{p}_t, \mathbf{p}_t \neq \mathbf{p}_u$, where $\operatorname{atan2}(y, x)$ is a function that yields an angle $\theta \in [-\pi, \pi)$ between a point (x, y) and the positive x-axis. The distance is given by $r = \|\overrightarrow{\mathbf{p}_t\mathbf{p}_u}\|_2$. Figure 4.4 (left) illustrates how the mapping ψ processes data triplets to derive the desired position $\mathbf{p}$.

By iterating through a trajectory, we obtain a set $\mathcal{S}_{t_\Delta}$ of positions $\mathbf{p}$ for a given time horizon t_Δ in the local coordinate system of a player. As argued earlier, this set should be constrained on the current velocity of the player. Hence, we obtain multiple sets, one for each velocity range. Let $\mathcal{V}$ be a partition of the speed range which is achievable by humans, e.g., $[0, 45]$, then, for every disjoint subset $V \in \mathcal{V}$ we obtain a set of points $\mathcal{S}_{t_\Delta, V} = \{(\mathbf{p}, v) \mid v \in V\} \subseteq \mathcal{S}_{t_\Delta}$. This procedure is outlined in Algorithm 1.

4.4. ZONES OF CONTROL

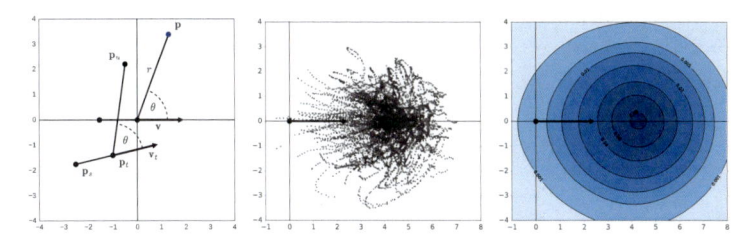

Figure 4.4: Illustration of the ψ function (left). The centre figure visualises the point distribution of a player, which is the set $\mathcal{S}_{t_\Delta,V}$, for time window $t_\Delta = 1s$ and velocity $14-20\ km/h$. The curves in the figure result from the player moving along a trajectory while maintaining a velocity within $14-20\ km/h$. Applying a KDE with a Gaussian kernel to the set $\mathcal{S}_{t_\Delta,V}$ then yields the movement model (right).

Finally, a probability distribution can be learned for every set $\mathcal{S}_{t_\Delta,V}$. The distribution might be discrete, defined on a finite grid in a two-dimensional space similar to the heat maps explained above. Another approach is to assume a distribution and estimate its parameters using a maximum likelihood-based approach. In this chapter we model the distribution using a kernel density estimate (KDE) (Parzen, 1962; Rosenblatt, 1956). The likelihood of the player reaching position $\mathbf{p}$ within a time horizon t_Δ given a current velocity of v_t, current position $\mathbf{p}_t$ and a previous position $\mathbf{p}_{t-t_\delta}$ is given by

$$\mathcal{P}_{t_\Delta}(\mathbf{p} \mid \mathbf{p}_t, \mathbf{p}_{t-t_\delta}, v_t) = \mathcal{P}^{\text{KDE}}_{t_\Delta,V}\left(\psi(\mathbf{p}_{t-t_\delta}, \mathbf{p}_t, \mathbf{p})\right),$$

where the KDE is evaluated on $\mathcal{S}_{t_\Delta,V}$ and $v_t \in V$ holds. An example of the set $\mathcal{S}_{t_\Delta,V}$ as well as the corresponding KDE is depicted in Figure 4.4 (center and right).

Figure 4.5 shows a visualisation of the movement models. The rows show the velocities of the players, while the columns encode the time horizons. The distribution of whereabouts at the end of the time horizons is then shown for every combination.

4.4 Zones of Control

The movement models themselves serve more as intermediate results as they are difficult to interpret and analyse. Yet, they can be used to compute zones of control (or dominant regions) per player or team (Brefeld et al., 2019; Gudmundsson and Wolle, 2014; Horton

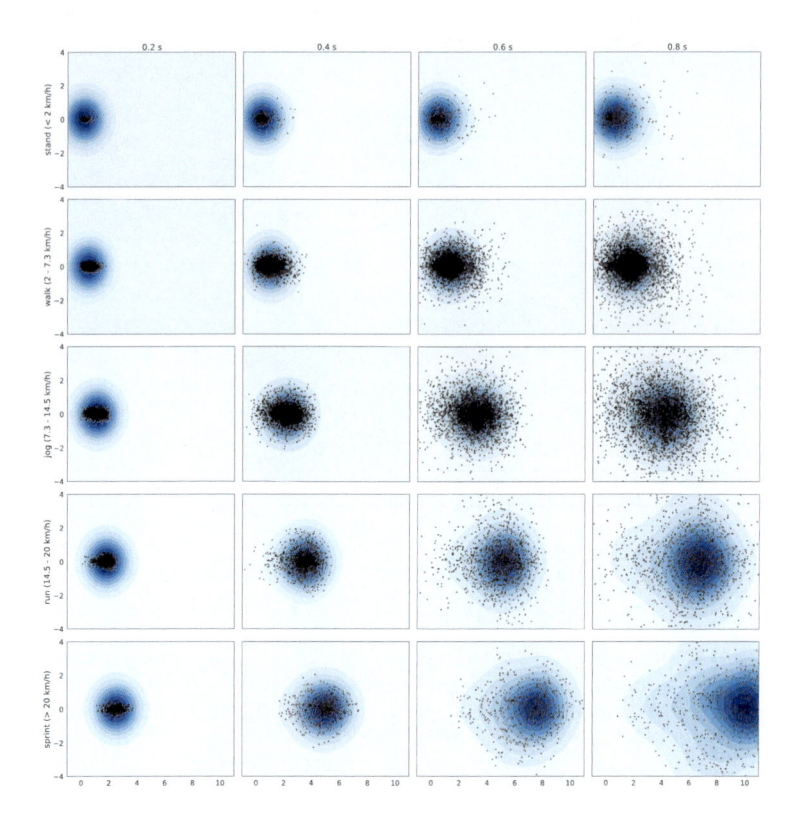

Figure 4.5: Visualisation of the movement model for different player initial speed values and time horizons.

et al., 2015; Spearman et al., 2017; Taki and Hasegawa, 2000). A player's zone of control is defined as follows.

Definition 1 (Taki and Hasegawa 2000) *Let F be the subset of $\mathbb{R}^2$ representing the pitch. The zone of control of player i is defined as the subset $D^i \subseteq F$, where the player i can arrive before any other player $j \neq i$.*

Formally, this can be expressed as follows. Let $\Gamma : \mathbb{R}^2 \times \mathbb{R}^2 \times \mathbb{R} \to \mathbb{R}_{\geq 0}$, $(\mathbf{p}, \mathbf{p}_t, v_t) \mapsto s$ be the function which yields the time s needed for player k to reach the position $\mathbf{p}$ given that her/his current position is $\mathbf{p}_t^k$ and her/his current velocity is v_t^k, i.e., $\Gamma(\mathbf{p} \mid \mathbf{p}_t^k, v_t^k) = s$. Then,

4.4. ZONES OF CONTROL

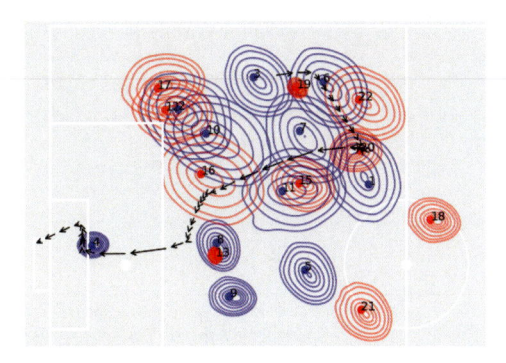

Figure 4.6: A section of a football field with players in red and blue, as well as their movement models for a fixed time horizon. Arrows indicate the movement of the ball.

Algorithm 2 Finite approximation

Input: Movement models $\mathcal{P}_{t_\Delta}^k$ for players, finite grid G
Output: Set B containing dominant player for every position
1: $B = \emptyset$
2: **for** $\mathbf{g} \in G$ **do**
3: $\quad B = B \cup \{(\mathbf{g}, \phi_{t_\Delta}(\mathbf{g}))\}$
4: **end for**

---

the set $D^i \subseteq F$ is defined such that for all $\mathbf{p} \in D^i$

$$i = \underset{k \in \{1,2,\ldots,K\}}{\arg\min} \Gamma(\mathbf{p} \mid \mathbf{p}_t^k, v_t^k),$$

where $\mathbf{p}_t^k$, v_t^k denote the position and velocity of player k at time t, and K is the total number of players. The zone of control of a team is defined as the union of sets D^i for its players.

Since we focus on a probabilistic view on this problem, we replace time, given by function Γ, with likelihood, derived from the movement models of the previous section for a given time horizon. Figure 4.6 shows an example. Computation of the resulting zones of control is similar and described below.

Let $\mathcal{P}_{t_\Delta}^k(\mathbf{p} \mid \mathbf{p}_t^k, \mathbf{p}_{t-t_\delta}^k, v_t^k)$ be the movement model of the k-th player as introduced in the previous section. It quantifies the likelihood of player k to reach position $\mathbf{p}$ given her/his current position $\mathbf{p}_t^k$ and last position $\mathbf{p}_{t-t_\delta}^k$, her/his current velocity v_t^k, and the time

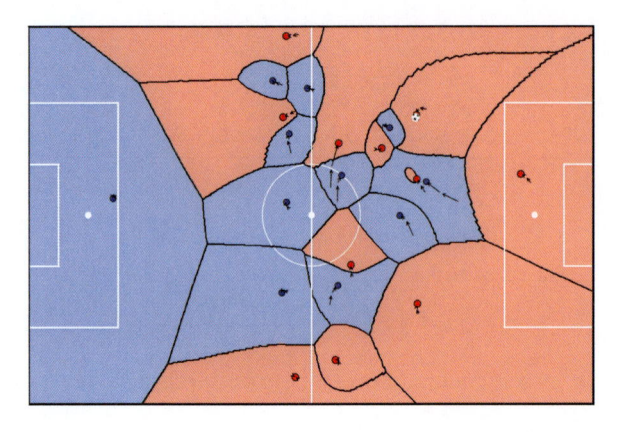

Figure 4.7: Zones of control. The arrows indicate the movements of the last two seconds.

horizon t_Δ. A position **p** is controlled by the player who has the highest likelihood according to her/his movement model; let ϕ_{t_Δ} be the function returning the index of this player. This function is defined as

$$\phi_{t_\Delta} : F \to \{1, 2, \ldots, K\}, \quad \mathbf{p} \mapsto \underset{k \in \{1,2,\ldots,K\}}{\arg\max} \; \mathcal{P}_{t_\Delta}^k \left(\mathbf{p} \mid \mathbf{p}_t^k, \mathbf{p}_{t-t_\delta}^k, v_t^k \right).$$

The zone of control of player i is given as the set $D^i = \{\mathbf{p} \in F \mid \phi_{t_\Delta}(\mathbf{p}) = i\}$ of all points that are controlled by her/him. If ties are broken, the set $\{D^1, D^2, \ldots, D^K\}$ is a partition of F.

A direct application of ϕ_{t_Δ} on all points of F is infeasible because F is not iterable since it is uncountable. A remedy is to use a finite approximation of the playing area. Let $G \subset F$ be a finite grid over F containing $n_x \cdot n_y$ equally spaced points in F with (axis-aligned) distance ε to each other. The zones of control are then computed on G rather than F. This way we obtain a finite approximation with precision ε. The procedure is summarised in Algorithm 2. For visualisation purposes, the set $B = \{(\mathbf{g}, \phi_{t_\Delta}(\mathbf{g})) \mid \mathbf{g} \in G\}$ can then be used to compute zones of control by assigning each position $\mathbf{p} \in F$ the same label as its closest neighbour from the grid G. An example is shown in Figure 4.7.

4.5. INTELLIGENT RETRIEVAL

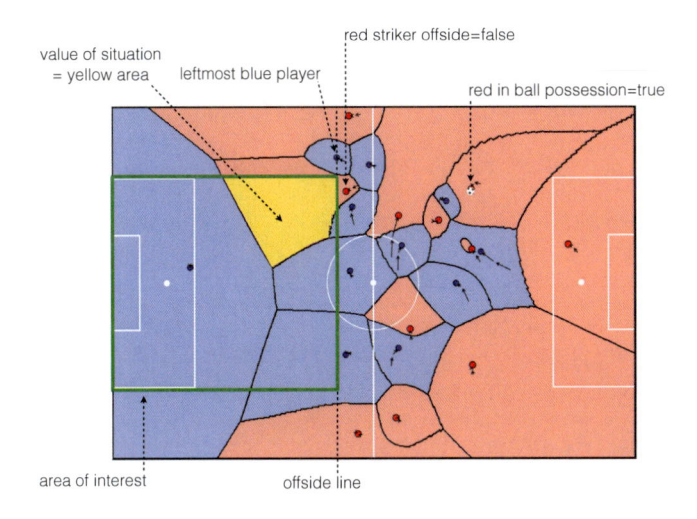

Figure 4.8: Retrieval of interesting situations.

4.5 Intelligent Retrieval

Retrieval of basic events like passes, shots, set pieces, etc. can be realised with the event log that accompanies the positional data. The event log annotates a multitude of basic events which can be directly incorporated into a database for retrieval.

When it comes to intelligent retrieval, we need to resort to positional data again. Intelligent retrieval is the task to find interesting and possibly complex situations that match a query. Consider the following example: A video analyst aims to, for example, find situations where a particular striker *could* have achieved a one-on-one with the goalkeeper in a central position. Depending on the results, she/he could report actionable insights back to the coach, e.g., the ball possessing player did not see the striker, the pass came too late and the striker was already offsides, etc. To complete the task, she/he would need to go through the entire video footage of historic games to collect the matching scenes.

By contrast, ideas from controlled zones can be leveraged to lift computer-based retrieval to the next level. Figure 4.8 shows the same situation as Figure 4.7 with additional annotations. Assuming a logic-based query language, the task of the video analyst could be formulated in terms of filters as follows. The first filter asks for ball possession. That is, the red team needs to have ball possession, otherwise the red striker cannot stage an attack. A second filter identifies the offside line based on the position of the leftmost blue player and asks whether the red striker is offside. A third filter then creates a central area of interest, defined by the box and the offside line. Only those situations happening within the area of interest are considered. The value of the situation is then the intersection of the zone of control of the red striker and the area of interest. If the intersection is empty, the situation is ignored. However, as shown in the figure, whenever the striker creates space behind the blue defenders, the intersection is not zero and the respective area (shown in yellow) is returned as the value of the situation.

The video analyst can either enter the query step-by-step or combine the filters to a complex query by logical *and* operations. The process can be trivially parallelised to deliver a ranking in only small amounts of time. The final ranking is sorted according to the value of the situations, that is, the scene that realises the largest overlap with the area of interest is ranked first, followed by the scene with the second largest overlap and so on. This way, the video analyst has access to the most relevant situations and can quickly jump from situation to situation.

The example sketches a novel way to combine data analysis with traditional video-based approaches. The application can be extended to pressing, counterattacks, and other interesting situations.

4.6 Conclusions

In this chapter we introduced different ways to process positional data from football matches. We began with simple position-based analyses like heat maps. Extracting velocities and accelerations then allowed to compute movement models. These again form the basis to derive zones of control. The final part of the chapter discussed a novel application in the context of intelligent retrieval.

Acknowledgments

The authors would like to thank Hendrik Weber and DFL / Sportec Solutions for providing the positional data for this study.

Bibliography

Bornn, L., Cervone, D., and Fernandez, J. (2018). Soccer analytics: Unravelling the complexity of "the beautiful game". *Significance*, 15(3):26–29.

Brefeld, U., Lasek, J., and Mair, S. (2019). Probabilistic movement models and zones of control. *Machine Learning*, 108(1):127–147.

Fonseca, S., Milho, J., Travassos, B., and Araújo, D. (2012). Spatial dynamics of team sports exposed by Voronoi diagrams. *Human Movement Science*, 31(6):1652–1659.

Fujimura, A. and Sugihara, K. (2005). Geometric analysis and quantitative evaluation of sport teamwork. *Systems and Computers in Japan*, 36(6):49–58.

Gudmundsson, J. and Wolle, T. (2014). Football analysis using spatio-temporal tools. *Computers, Environment and Urban Systems*, 47:16–27.

Haase, J. and Brefeld, U. (2014). Mining positional data streams. In *International Workshop on New Frontiers in Mining Complex Patterns*, pages 102–116. Springer.

Harmon, M., Lucey, P., and Klabjan, D. (2016). Predicting shot making in basketball learnt from adversarial multiagent trajectories. *ArXiv e-prints*.

Horton, M., Gudmundsson, J., Chawla, S., and Estephan, J. (2015). Automated classification of passing in football. In *Pacific-Asia Conference on Knowledge Discovery and Data Mining*, pages 319–330. Springer.

Manafifard, M., Ebadi, H., and Moghaddam, H. A. (2017). A survey on player tracking in soccer videos. *Computer Vision and Image Understanding*, 159:19–46. Computer Vision in Sports.

Parzen, E. (1962). On estimation of a probability density function and mode. *The Annals of Mathematical Statistics*, 33(3):1065–1076.

Pettersen, S. A., Johansen, D., Johansen, H., Berg-Johansen, V., Gaddam, V. R., Mortensen, A., Langseth, R., Griwodz, C., Stensland, H. K., and Halvorsen, P. (2014). Soccer video and player position dataset. In *Proceedings of the 5th ACM Multimedia Systems Conference*, MMSys '14, pages 18–23, New York, USA. ACM.

Rosenblatt, M. (1956). Remarks on some nonparametric estimates of a density function. *The Annals of Mathematical Statistics*, 27(3):832–837.

Spearman, W., Pop, P., Basye, A., Hotovy, R., and Dick, G. (2017). Physics-based modeling of pass probabilities in soccer. In *Proceedings of the 11th MIT Sloan Sports Analytics Conference*, pages 1–14.

Taki, T. and Hasegawa, J. (2000). Visualization of dominant region in team games and its application to teamwork analysis. In *Proceedings of the International Conference on Computer Graphics*, CGI '00, pages 227–235, Washington, DC, USA. IEEE Computer Society.

Taki, T., Hasegawa, J., and Fukumura, T. (1996). Development of motion analysis system for quantitative evaluation of teamwork in soccer games. In *Proceedings of 3rd IEEE International Conference on Image Processing*, pages 815–818.

Voronoi, G. (1908). Nouvelles applications des paramètres continus à la théorie des formes quadratiques. Premier mémoire. Sur quelques propriétés des formes quadratiques positives parfaites. *Journal für die Reine und Angewandte Mathematik*, 133:97–178.

Zhao, Y., Yin, F., Gunnarsson, F., Hultkratz, F., and Fagerlind, J. (2016). Gaussian processes for flow modeling and prediction of positioned trajectories evaluated with sports data. In *2016 19th International Conference on Information Fusion (FUSION)*, pages 1461–1468.

Chapter 5

Ranking and Prediction Models for Football Data

ANDREAS GROLL
TECHNISCHE UNIVERSITÄT DORTMUND

GUNTHER SCHAUBERGER
TECHNISCHE UNIVERSITÄT MÜNCHEN

HANS VAN EETVELDE
GHENT UNIVERSITY

Abstract

In this chapter we present the most common methods in ranking of teams and prediction of matches in association football. The main ranking-methods can be categorised in point-winning systems, least squares methods, maximum likelihood-based methods, and Elo ratings. We explain the idea and operation of these methods, possible extensions, and their use in the current and former world of association football. Furthermore, we present and illustrate major approaches for the modelling and prediction of football matches. In principal, two main approaches can be distinguished, namely the prediction of the scores of both competing teams in a football match and the prediction of the match outcomes represented by the three categories *win*, *draw* and *loss*. The key steps of these strategies are presented and illustrated via real data examples. Finally, some extensions and further developments are outlined.

5.1 Introduction

Football has always been extremely popular around the world, both among fans and among the many people who play this sport themselves. In many countries football is the number-one sport, and the World Cup Final has been one of the biggest TV events in the world for decades, with more than a billion people following the final in Moscow at the 2018 World Cup in Russia on the screens according to the FIFA[1].

There are many questions that affect countless football fans week by week when they cheer for their club either in the stadium or in front of the TV. Which team is the strongest team in the competition? Does the home team have a higher chance to win a football match? Can a match be decided by the players' mileage? How much information is contained in the betting odds of bookmakers? And there are many more. Some of these questions will be addressed in this chapter from a statistical perspective.

Generally, among the statistical community, the modelling and prediction of football matches has also gained increasing interest in the last years and has become a major research area within the field of sports statistics (see Albert et al., 2005). Mainly, one can distinguish two basic modelling strategies for football data, the *ranking methods* and *covariate-based methods*. While the former methods are mostly based on historic match results, the latter are regression-type approaches including covariate information and typically have a high capacity with respect to prediction of new matches.

Principally, ranking-methods can be categorised in point-winning systems, least squares methods, maximum likelihood-based methods, and Elo ratings. In this chapter, we explain the basic idea and operation of these ranking methods, and discuss possible extensions and their use in current and former association football. The covariate-based regression approaches can mainly be divided into two model classes. The first class considers models for the precise scores of both competing teams in a match, while the second approach is to model the coarser match outcome, usually measured by the categories *win*, *draw* and *loss*. In both cases a major goal of the regression approaches is to provide predictions of future matches. Here, the main idea is to learn covariate effects from past matches by fitting a statistical model and to use these for future matches.

For both ranking and regression methods different general aspects

---

[1] https://www.fifa.com/worldcup/news/more-than-half-the-world-watched-record-breaking-2018-world-cup

need to be adequately considered. For example, several differences exist between models for national football leagues and models for international tournaments like the FIFA World Cup or the UEFA European Championship (EURO). Typically, in national leagues matches are played on the home ground of one of the competing teams. It is widely accepted that due to crowd support, absence of travel stress, referee bias, etc., the hosting team has a general advantage over its opponent, which makes it necessary to account for this so-called home effect. While home effects are less pronounced in international tournaments they typically play a considerable role in national leagues. Also, while leagues are mostly designed as so-called round-robins (mostly double round robins), international tournaments typically consist of a group stage followed by a knockout stage leading to one final to determine the tournament champion. Such aspects need to be considered when setting up an appropriate statistical ranking or prediction model.

The main purpose of the present chapter is to give an introduction into the basic concepts of ranking (Section 5.3) and prediction models (Section 5.4) for football and to give an overview over the most important approaches in this research area. In the first part we describe the most common probabilistic models used for football matches, which will have their application in both ranking and prediction.

5.2 Modelling of Football Matches

We first present the most frequently used modelling approach for precise scores. For this approach, the scores of both teams are assumed to be (conditionally independent) random variables, each following a Poisson distribution. Alternatively, also different distributions would be possible, for example, the negative binomial distribution or a bivariate Poisson distribution for both scores referring to the same match. However, for the following introduction we will restrict ourselves to the most popular (and pleasantly simple) Poisson distribution.

Alternatively, to model the aims of both teams in a football match directly, one can consider the classification into the categories *win*, *draw* and *loss*. These models are presented in Section 5.2.2.

5.2.1 Poisson Models

In the Poisson model the number of goals a team r scores in a match against opposing team s is assumed to be Poisson distributed, i.e.

$$Y_{rs} \sim Po(\lambda_{rs}),$$

and hence the probability for a specific number of goals $y \in \mathbb{N}_0$ is given by

$$P(Y_{rs} = y) = \frac{\lambda_{rs}^y \cdot \exp(-\lambda_{rs})}{y!}. \qquad (5.1)$$

Here $\lambda_{rs} \in \mathbb{R}_+$ is the corresponding (positive) Poisson event rate, controlling the intensity with which goals are scored by team r in a match against team s. More technically speaking, the parameter λ_{rs} represents the expected number of goals team r scores, i.e. $E[Y_{rs}] = \lambda_{rs}$, and to some extent also quantifies the uncertainty as it simultaneously reflects the corresponding variance, i.e. $Var(Y_{rs}) = \lambda_{rs}$. For this reason, the parameter λ_{rs} plays a crucial role when it comes to the incorporation of information on both competing teams. In the simplest case we assume that the goals team r scores in a match against opposing team s are affected by the playing skills of both teams, denoted by γ_r and γ_s, respectively, and by a potential home effect $\delta \in \mathbb{R}$ of one of the teams (if applicable). This results in the following structure of the Poisson event rate

$$\lambda_{rs} = \exp(\beta_0 + \delta \cdot \mathbb{1}(\text{team } r \text{ is home team}) + \gamma_r - \gamma_s), \qquad (5.2)$$

where $\eta_{rs} := \beta_0 + \delta \cdot \mathbb{1}(\text{team } r \text{ is home team}) + \gamma_r - \gamma_s$ is the so-called *linear predictor*. The exponential function $\exp(\cdot)$ in equation (5.2) guarantees that λ_{rs} is positive and the intercept β_0 accounts for the average number of goals. Moreover, we assume that the ability γ_r of team r has a positive effect on the number of goals team r is able to score, while the ability of the opposing team s counteracts and, hence, γ_s has a negative sign. Finally, $\mathbb{1}(\cdot)$ is an indicator function indicating whether the team whose goals are considered is playing at home.

Typically, given the home effect and the playing skills of both teams, the match scores are assumed to be independent, i.e.

$$P(Y_{rs} = y_{rs}, Y_{sr} = y_{sr}) = P(Y_{rs} = y_{rs}) \cdot P(Y_{sr} = y_{sr}). \qquad (5.3)$$

This, of course, is a rather strict assumption and there exist several extensions in the literature regarding the correlation structure of the two scores in a match. For example, Karlis and Ntzoufras (2003) introduced the idea to model the scores of both teams via a *bivariate Poisson distribution* which inherently can account for (positive) dependencies between the scores. This idea was then adopted in several subsequent publications (e.g., Koopman and Lit, 2015; Groll et al., 2018). While the bivariate Poisson distribution can only consider positive dependencies, *copula-based models* can also account for negative dependencies between the scores. For the modelling of football results, copula approaches were first proposed by McHale and Scarf (2007), later the idea was taken up, for example, by McHale and Scarf (2011), Boshnakov et al. (2017) and van der Wurp et al. (2020). However, for simplicity and clarity, in the present work we focus on the independent

5.2.2 Ordinal Models

In contrast to the Poisson models from Section 5.2.1 considering the scores of both teams, in ordinal regression methods each match is represented by a single-valued outcome: let for a certain match i the random variable $Z_{irs} \in \{1, 2, 3\}$ denote the ordinal match response, where $Z_{irs} = 1$ represents a win of the (first-named) home team r over team s, $Z_{irs} = 2$ a draw, and $Z_{irs} = 3$ an away win, respectively. The simplest model for such a response variable can be denoted by

$$P(Z_{irs} \leq k) = F(\delta + \theta_k + \gamma_r - \gamma_s), \quad k = 1, 2, 3, \qquad (5.4)$$

where F stands for the cumulative distribution function, γ_r and γ_s are again the ability parameters for teams r and s, δ represents a home effect parameter, and θ_k, $k = 1, 2, 3$ are threshold parameters for the separate categories. The threshold parameters determine the transition probabilities between the response categories, where the restriction $\theta_3 = \infty$ ensures that the cumulative probability for the last category equals one. Moreover, the threshold parameters have to be restricted with $\theta_1 = -\theta_2$ to guarantee for identifiability of all parameters[2]. Finally, $F(\cdot)$ represents a response function; typically the logistic distribution function $F(\cdot) = \exp(\cdot)/(1 + \exp(\cdot))$ or the standard normal distribution function $F(\cdot) = \Phi(\cdot)$ (probit model) are chosen. This model represents a cumulative regression model and can be seen as an extension of the classical binary regression model to ordinal responses (see, e.g., Tutz, 1986, and Agresti, 1992, for a general introduction to regression models for ordered responses, allowing for a general number of K categories). To get the probability masses of the outcomes, we can easily see that $P(Z_{irs} = 1) = P(Z_{irs} \leq 1)$, $P(Z_{irs} = 2) = P(Z_{irs} \leq 2) - P(Z_{irs} \leq 1)$ and $P(Z_{irs} = 3) = 1 - P(Z_{irs} \leq 2)$.

In many publications this model class is also referred to as Bradley-Terry models (Bradley and Terry, 1952), which are models for the analysis of paired comparisons. In paired comparisons, pairs of two objects out of a larger set of objects are compared with respect to a certain latent trait, mostly resulting in a binary outcome which represents the preference or dominance of one object over the other. Football matches can be treated as paired comparisons between the

---

[2] Without this restriction, θ_1 and θ_2 would be non-identifiable due to the presence of the home effect parameter δ. For any constant c, the sets $\{\delta^{(1)}, \theta_1^{(1)}, \theta_2^{(1)}\}$ and $\{\delta^{(2)}, \theta_1^{(2)}, \theta_2^{(2)}\}$, where $\{\delta^{(1)} = \delta^{(2)} + c, \theta_1^{(1)} = \theta_1^{(2)} - c, \theta_2^{(1)} = \theta_2^{(2)} - c\}$, would be equivalent solutions. Another (yet less intuitive) possibility would be to eliminate the home effect parameter δ, i.e. impose the restriction $\delta = 0$. Then, the model would still contain a "hidden" home effect, namely the mean of θ_1 and θ_2.

competing teams where the playing abilities are treated as the latent traits. In contrast to simple paired comparisons, in football matches draws between the objects are also possible, which automatically leads to ordinal instead of binary responses.

Several different versions of this basic model exist in the literature. A first version is presented in Fahrmeir and Tutz (1994), where the teams' abilities are allowed to vary over time. Other modifications and extensions of the model (5.4) are investigated, for example, in Koning (2000) and Knorr-Held (2000).

5.3 Ranking Methods

In association football, ranking is used for many different purposes. In national leagues the final table is used to determine the winner of the league as well as the teams which will be promoted to a higher division or which will be relegated to the lower division. For national teams, the FIFA ranking is used to decide which teams are in which pot in the draws for the World Cup and the classification groups. These are just a few examples of how important ranking is in football.

In this part we discuss several different methods for ranking teams based on match results. We start by shortly discussing the point-winning ranking system, since this is the most commonly used ranking method in football. Thereafter, we discuss more statistical ranking methods, which can be categorised in least squares ratings, maximum likelihood methods, and Elo-ratings.

5.3.1 Point-Winning Systems

In national football leagues the final ranking is mostly determined by a *"three points for a win"* system. Teams receive 3 points for winning a game and 1 point in case of a draw. Losing teams do not receive any points. At the end of the season, the ranking is made by sorting the teams by their total number of points. When teams have the same number of points, other criteria are used to assign the final ranking, e.g., overall goal difference, total number of goals scored, the results in mutual matches, etc. These rankings are not only used in leagues but also in early phases of big tournaments as the World Cup. Originally, most football competitions used the "two points for a win" system, but, in 1981, the Football Association in England decided to reward a win with three points, hoping that this would result in a more attractive playing style. In 1995 it was adopted by the FIFA and it became the standard in national competitions[3].

---

[3] https://www.questia.com/newspaper/1G1-136226360/how-three-points-for-a

5.3. RANKING METHODS

Point-winning systems are quite easy to understand and are a very fair way of rewarding when the teams play the same number of matches against similar opponents. For that reason, it perfectly suits the aim of the league tables, but it is not useful in competitions where teams do not all play against each other, e.g., if the purpose is to rank national teams.

5.3.2 Least Squares Models

The *least squares* method is a very standard approach in regression analysis. Stefani (1977) developed a least squares system for team ratings, using the margin of victory. In the least squares method, we assume that the difference in ratings is reflecting the average goal difference between two teams. Assume we have n matches of K teams with ratings $\gamma_1, ..., \gamma_K$. The goal difference d_i in match i, where teams r and s play against each other, is defined as follows:

$$d_i = \delta \cdot \mathbb{1}(\text{team } r \text{ is home team}) + \gamma_r - \gamma_s + e_i \quad \text{for } i = 1, ..., n,$$

where $\delta \in \mathbb{R}$ represents the home effect and e_i is the error term for match i. Now the ratings $\gamma_1, ..., \gamma_K$ are chosen such that the mean squared error $\frac{1}{n}\sum_{i=1}^{n} e_i^2$ is minimised.

The biggest advantage of the least squares method is the clear interpretation of the (difference in) ratings, which is lacking in all of the other proposed methods. If the purpose is to reflect the current strength of the teams, one could give more weight to recent matches. The expression that is optimised is then given by

$$\frac{1}{n}\sum_{i=1}^{n} w_i e_i^2,$$

where w_i is the weight assigned to match i.

5.3.3 Maximum Likelihood Methods

For the *maximum likelihood* methods, we use the modelling approaches that were discussed in Section 5.2. We consider γ_r and γ_s in equations (5.2) and (5.4) as parameters which should be estimated. The values of these parameters that return the highest likelihood are the estimated team strengths. As in Section 5.2, we will first discuss the Poisson models, followed by the ordinal models.

-win-has-fouled-up-football

Poisson Ranking Method

Assume we have n past match results and say y_{irs} is the number of goals scored by team r against team s in match i. The likelihood function is then given by

$$\prod_{i=1}^{n} P(Y_{irs} = y_{irs}) \cdot P(Y_{isr} = y_{isr}).$$

These probabilities are given by the Poisson probability mass function (5.1), where the Poisson parameters λ_{irs} and λ_{isr} depend on the team strength parameters via equation (5.2) such that the likelihood function depends on the team strengths of both teams γ_s and γ_r. If we consider in total K teams, the maximum likelihood procedure will estimate $K+2$ parameters, namely the strengths $\gamma_1, ..., \gamma_K$, the intercept parameter β_0, and the home effect parameter δ. Instead of the independent Poisson model, one could also choose other count distributions and dependency structures which were mentioned at the end of Section 5.2.1.

Ordinal Outcome

Instead of the Poisson model, we can also use the ordinal outcome models to construct a likelihood function. Let z_{irs} be the result of match i between teams r and s: $z_{irs} = 1$ represents a win for the home team r in match i, $z_{irs} = 2$ represents a draw in match i, and $z_{irs} = 3$ represents a win for the away team s. If we again consider a total of n matches, the likelihood function will be given by

$$\prod_{i=1}^{n} P(Z_{irs} = z_{irs}).$$

As was mentioned before, we set $\theta_3 = \infty$, and we assume that $\theta_1 = -\theta_2$. Therefore, we again have to estimate $K+2$ parameters, namely the team strengths $\gamma_1, ..., \gamma_K$, the threshold parameter θ_1, and the home effect parameter δ.

Weighted Maximum Likelihood

The maximum likelihood method has some interesting extensions. One possibility is to add weights to the matches. If we want to have a ranking that reflects the current strength of a team rather than the average strength over a period of time, we could give more weight to recent matches. A good choice would be to add an exponential weight function depending on the date of the match played given by the formula

$$w_{t_i} = \exp(-\alpha t_i), \tag{5.5}$$

5.3. RANKING METHODS

where t_i represents the number of days ago that match i took place, and α is a parameter that determines the slope of the curve. The weighted likelihood function for the Poisson model would then be given by

$$\prod_{i=1}^{n} \Big(P(Y_{irs} = y_{irs}) \cdot P(Y_{isr} = y_{isr}) \Big)^{w_{t_i}}.$$

Another possible extension is to add weights according to the importance of a game. Weighted maximum likelihood methods have proven to be effective in ranking teams according to their current strengths, and so they are particularly useful for prediction. Ley et al. (2019) have compared different weighted maximum likelihood ranking methods in terms of their predictive performance. They find that the rankings based on the number of goals, the independent and bivariate Poisson models, perform better than the ordinal models, namely the Bradley-Terry (based on the logistic distribution) and Thurstone-Mosteller (based on the normal distribution) models.

5.3.4 Elo Models

The Elo-ratings system was invented by Elo (1978) to serve as a new ranking method in chess, but it is nowadays used in many other sports. The idea is to update the rating of a team after each match according to an update formula which depends on the expected outcome of the match and the true outcome of the match. If a team performs better than expected, its rating will rise, if it performs worse, its rating will decrease. The update formula is the following:

$$\gamma_{r,\text{new}} = \gamma_r + I * (W - W_e),$$

where

- $\gamma_{r,\text{new}}$ stands for the new number of points or rating of team r and γ_r is its rating before the match was played.

- I is the update factor which determines the smoothness of the ranking. If you choose I too small, this will lead to a ranking that is very smooth, but that is not reflecting the current strength of the teams. A large value of the factor I will give too much importance to recent matches and so will lead to sudden changes in the ranking.

- W reflects the result of a match, $W = 1$ in case of a win, $W = 0.5$ in case of a draw and $W = 0$ when you lose.

- W_e stands for the expected performance and is calculated as the probability of winning if there were only two possibilities: winning or losing. It is based on the Bradley-Terry model using the logistic distribution, which was discussed in Section 5.2.2. Assume that we only have two possible outcomes: winning and losing a game. Using the logistic distribution function, equation (5.4) results in

$$W_e = P(\text{``Win''}) = \frac{1}{\exp(-(\delta + \gamma_r - \gamma_s)/s) + 1}, \quad (5.6)$$

where $s > 0$ is a scaling parameter, which can be chosen arbitrarily.

Both the FIFA/Coca-Cola World Ranking for men and for women are examples of Elo-systems, but there are some differences. The women's ranking makes use of the home effect parameter δ, while the ranking for men does not. In the women's ranking the performance W is also based on the number of goals scored and the goal difference, while in the men's ranking it is solely based on the outcome of the match. The description of both procedures can be found on https://www.fifa.com/fifa-world-ranking/procedure/.

Thanks to its statistical basis, the Elo-system seems to have good predictive properties, as is shown by Hvattum and Arntzen (2010) and Lasek et al. (2013). The advantage of the Elo-rating compared with the maximum likelihood method is that the updating of the rating is very straightforward, as is shown in the second example in the next subsection.

5.3.5 Examples

German Bundesliga: Point-Winning and Least Squares

In this example we apply the least squares method to the matches of the German Bundesliga in the season 2018-2019 and compare it with the official ranking, which is based on a point-winning system. The result is shown in Table 5.1. We see that the two rankings are highly correlated, with a Spearman correlation coefficient of $\rho_S = 0.96$. Both have Bayern Munich as the best team and Nurnberg as the worst team in the competition. The advantage of the least squares method is that we can now give an interpretation to the ratings, e.g., we see that the rating of Bayern Munich is 1.38 and the rating of Wolfsburg is 0.35. This means that we expect a goal difference of 1.03 in favour of Bayern Munich when those teams would face each other.

National Teams: FIFA/Elo and Maximum Likelihood

In this part we compare the official FIFA-ranking for men, based on the Elo-system, with a weighted maximum likelihood method, applied

5.3. RANKING METHODS

Table 5.1: Ranking for the German Bundesliga in the season 2018-2019, based on the least squares system and the point-winning system.

Ranking	Team	Least Squares Rating	Team	Points
1	Bayern Munich	1.38	Bayern Munich	78
2	Dortmund	1.24	Dortmund	76
3	RB Leipzig	1.06	RB Leipzig	66
4	Hoffenheim	0.57	Leverkusen	58
5	Wolfsburg	0.35	M'gladbach	55
6	Leverkusen	0.32	Wolfsburg	55
7	M'gladbach	0.28	Ein Frankfurt	54
8	Ein Frankfurt	0.15	Werder Bremen	53
9	Fortuna Düsseldorf	0.09	Hoffenheim	51
10	Werder Bremen	0.02	Fortuna Düsseldorf	44
11	Hertha	-0.06	Hertha	43
12	Freiburg	-0.29	Mainz	43
13	Mainz	-0.40	Freiburg	36
14	Schalke 04	-0.47	Schalke 04	33
15	Augsburg	-0.64	Augsburg	32
16	Hanover	-1.11	Stuttgart	28
17	Stuttgart	-1.15	Hanover	21
18	Nurnberg	-1.34	Nurnberg	19

on the national teams. The date of both rankings is 16th August 2018, so shortly after the FIFA World Cup in Russia. Again, both rankings are highly correlated ($\rho_S = 0.94$)

As an example, here we discuss how the ratings of France and Germany were updated after their match on the 6th of September 2018 in the UEFA Nations League. The result of the match was 0-0. For both teams, the performance measure W is equal to 0.5, since the match was a draw. The expected performance measure for France was

$$W_e = \frac{1}{10^{-(\gamma_F - \gamma_G)/600} + 1} = \frac{1}{10^{-(1726-1561)/600} + 1} = 0.65\,.$$

Note that the parameter s in equation (5.6) is chosen to be $600/\log(10)$. The update formula now gives us

$$P = 1726 + 15(0.5 - 0.65) = 1723.70\,,$$

where the importance factor I is chosen to be 15 for Nations League matches, according to the official FIFA ranking procedure. For Germany, we have

$$W_e = \frac{1}{10^{-(1561-1726)/600} + 1} = 0.35$$

and consequently

$$P = 1561 + 15(0.5 - 0.35) = 1563.30\,.$$

France as favourite loses some points since it performed worse than expected, while Germany on the other side was the underdog according to the FIFA ranking, and so it performs better than expected by tying against France. This results in an increased number of points.

On the right side of Table 5.2, we show the weighted maximum likelihood ratings using the Poisson model, with an exponential time weighting function as suggested in equation (5.5), since this method has been proven to have a good predictive performance. To make sense of these ratings, one should also have the estimates for the parameters β_0 and the home effect δ from equation (5.2). These estimates were found to be $\beta_0 = 0.0036$ and $\delta = 0.291$. Now, if we have, for example, Brazil playing at home against Spain, the expected number of goals for Brazil will be $\lambda_B = \exp(0.0036 + 0.291 + 1.46 - 1.31) = 1.56$, while the expected number of goals for Spain will be $\lambda_S = \exp(0.0036 + 1.31 - 1.46) = 0.86$. Given these expected number of goals, one can also deduce the probabilities for a home win, a draw, or an away win.

Table 5.2: Rankings for the national teams. On the left the official FIFA points based on the Elo-system are shown. On the right the ratings based on a Poisson maximum likelihood method are shown.

Ranking	Team	FIFA Points (Elo)	Team	Poisson Ratings
1	France	1726	Brazil	1.46
2	Belgium	1723	Spain	1.31
3	Brazil	1657	France	1.19
4	Croatia	1643	Belgium	1.17
5	Uruguay	1627	Germany	1.17
6	England	1615	Argentina	1.14
7	Portugal	1599	Colombia	1.12
8	Switzerland	1597	Netherlands	1.10
9	Denmark	1580	England	1.07
10	Spain	1580	Portugal	1.05
11	Argentina	1574	Chile	1.04
12	Chile	1570	Uruguay	1.02
13	Sweden	1565	Croatia	0.97
14	Colombia	1563	Peru	0.93
15	Germany	1561	Sweden	0.93
16	Mexico	1560	Italy	0.90
17	Netherlands	1540	Denmark	0.90
18	Poland	1538	Switzerland	0.88
19	Wales	1536	Ecuador	0.87
20	Peru	1535	Ukraine	0.85
21	Italy	1532	Poland	0.84
22	USA	1508	Mexico	0.84
23	Austria	1502	Serbia	0.82
24	Tunisia	1498	Russia	0.81
25	Senegal	1498	Bosnia and Herzegovina	0.81
26	Slovakia	1493	Morocco	0.77
27	Northern Ireland	1492	Turkey	0.77
28	Romania	1490	Romania	0.77
29	Republic of Ireland	1484	Austria	0.76
30	Paraguay	1477	Republic of Ireland	0.75

5.4 Prediction Models

The third part of this chapter is devoted to *covariate-based regression approaches*. A major goal of these approaches is to provide predictions of future football matches. Basically, the regression approaches can be divided into two model classes, depending on the type of the response variable. The first class consists of models for the precise match scores; the second class consists of models for the coarser categorical match outcome with categories *win*, *draw*, and *loss* (from the perspective of the first-named team, which usually is the home team).

5.4.1 Poisson Regression

Inclusion of Covariates

We will show how covariate information can be included in the basic Poisson model from Section 5.2.1. Let $i = 1, \ldots, n$ be football matches in a league or tournament with K different participating teams. We extend the linear predictor from equation (5.2) by allowing the two teams' abilities to depend on covariates. For this purpose, we define the vector $\mathbf{x}_{ir} = (x_{ir1}, \ldots, x_{irp})^T$, which collects p covariates corresponding to team r in match i. It could include variables such as the team's average age, market value, Elo rating, etc., which, principally, can vary over the matches of team r, making it necessary to use an additional index i to identify the regarded match. Analogously, $\mathbf{x}_{is}$ collects the same set of variables for the opposing team s in match i. We then obtain

$$\lambda_{irs} = \exp(\beta_0 + \delta \cdot \mathbb{1}(\text{team } r \text{ is home team in match } i) + \gamma_{ir} - \gamma_{is}), \quad (5.7)$$

with

$$\gamma_{ir} = \mathbf{x}_{ir}^T \boldsymbol{\beta} \quad \text{and} \quad \gamma_{is} = \mathbf{x}_{is}^T \boldsymbol{\beta}.$$

The vector $\boldsymbol{\beta} = (\beta_1, \ldots, \beta_p)^T$ collects the corresponding p regression effects of the covariates. These regression coefficients are currently assumed to be equal for both teams, which allows to display the linear predictor from model (5.7) in the more compact form

$$\eta_{irs} = \beta_0 + \delta \cdot \mathbb{1}(\text{team } r \text{ is home team in match } i) + (\mathbf{x}_{ir} - \mathbf{x}_{is})^T \boldsymbol{\beta}.$$

Hence, $\boldsymbol{\beta}$ actually corresponds to the regression effects of covariate differences between both competing teams (at this point implicitly assuming that these differences make sense, which is particularly valid for metric predictors). This basic predictor structure has been employed, e.g., in the regression approaches used in Schauberger and

Groll (2018) on FIFA World Cup data. Model (5.7) could be extended to, additionally to covariate effects, account for team-specific ability parameters β_{r0}, β_{s0}, yielding the linear predictor

$$\eta_{irs} = \beta_0 + \delta \cdot \mathbb{1}(\text{team } r \text{ is home team in match } i)$$
$$+ \beta_{r0} - \beta_{s0} + (\boldsymbol{x}_{ir}^T - \boldsymbol{x}_{is}^T)\boldsymbol{\beta}. \quad (5.8)$$

These team-specific abilities β_{r0}, β_{s0} principally could be incorporated as either fixed or random effects. Also, the abilities could be further decomposed into a separate attack and defense ability for each team, i.e. in a match between team r and team s the corresponding linear predictors then have the form

$$\eta_{irs} = \beta_0 + \delta \cdot \mathbb{1}(\text{team } r \text{ is home team in match } i)$$
$$+ \beta_{r0}^{att} - \beta_{s0}^{def} + (\boldsymbol{x}_{ir}^T - \boldsymbol{x}_{is}^T)\boldsymbol{\beta}$$

$$\eta_{isr} = \beta_0 + \delta \cdot \mathbb{1}(\text{team } s \text{ is home team in match } i)$$
$$+ \beta_{s0}^{att} - \beta_{r0}^{def} + (\boldsymbol{x}_{is}^T - \boldsymbol{x}_{ir}^T)\boldsymbol{\beta} \quad (5.9)$$

and, consequently, we get the two Poisson distributions $Y_{irs} \sim Po(\eta_{irs})$ and $Y_{isr} \sim Po(\eta_{isr})$ for the two scores of this match. This model formed the basis for the approach proposed in Groll et al. (2015) to predict the FIFA World Cup 2014 in Brazil.

If the assumption of equal regression coefficients for the covariates of the team, whose goals are considered (here: team r), and the opposing team (here: team s) are unrealistic or need to be relaxed, the model could be further generalised. Moreover, the home effect δ does not necessarily have to be equal for all teams and, instead, could vary over teams. This could be justified, for example, if stadium capacities or fan support substantially differ across teams. The corresponding predictor then yields

$$\lambda_{irs} = \exp(\beta_0 + \delta_r \cdot \mathbb{1}(\text{team } r \text{ is home team in match } i)$$
$$+ \beta_{r0} - \beta_{s0} + \boldsymbol{x}_{ir}^T \boldsymbol{\alpha} - \boldsymbol{x}_{is}^T \boldsymbol{\beta}). \quad (5.10)$$

Now, team-specific home effect parameters $\delta_k, k = 1, \ldots, K$, and different covariate effects $\boldsymbol{\alpha}$ and $\boldsymbol{\beta}$ for the two opposing teams will be obtained, increasing the overall number of parameters that need to be estimated (typically substantially) and, hence, the computational complexity when fitting the model. Moreover, it gets harder

5.4. PREDICTION MODELS

to interpret the model due to the large number of coefficients. In the following, again for simplicity and clarity, we focus on the simpler model (5.8). In particular, we will fit model (5.8) to match data from the German Bundesliga for the four seasons 2014/15–2017/18 in Section 5.4.4 and show the model's predictive performance by using a specific rolling window strategy.

Basically, all models (5.2)-(5.9) considered so far can be estimated within the framework of Poisson generalised linear models (GLMs). The different predictor structures need to be transferred into suitable design matrices and then standard software fitting routines for GLMs can be used. In particular, we used the `glm()` function from the statistical software program R (R Development Core Team, 2019) throughout this text when fitting Poisson models.

In case the team-specific abilities β_{r0}, β_{s0} in model (5.8) are treated as random effects, e.g., the `glmer()` function from the R package lme4 (Bates et al., 2015, 2017) can be used. Furthermore, if there are many covariates included in the model, techniques that enforce variable selection may be desirable. A popular approach in this regard is the Least Absolute Shrinkage Selection Operator (LASSO; Tibshirani, 1996), which penalises the absolute values of the regression coefficients to obtain a sparse model and a good predictive performance. The *LASSO* technique is implemented in the `glmnet` package (Friedman et al., 2010) in R and has been employed by Groll et al. (2015) in their prediction model for the FIFA World Cup 2014 in Brazil. Alternatively, so-called *boosting* methods could be beneficial. Boosting stems from machine learning and merges several weak learners into a single strong learner, this way achieving a good trade-off between bias and variance and a good capacity for prediction. A prominent implementation of boosting can be found in the `mboost` package (Hothorn et al., 2017) in R. This methodology was used, e.g., by Groll et al. (2018) to model UEFA EURO data.

Simulation and Prediction

One major advantage of the Poisson models from the previous section is that it is rather straightforward to draw exact match outcomes for single matches. If we have fitted our Poisson model to some historic match data and have obtained predictions $\hat{\lambda}_{irs}$ and $\hat{\lambda}_{isr}$ for the Poisson event rates of the two teams competing in a (new) match i, we can directly draw their goals from the corresponding predicted Poisson distributions, i.e., $G_{irs} \sim Poisson(\hat{\lambda}_{irs})$, $G_{isr} \sim Poisson(\hat{\lambda}_{isr})$. Here, G_{irs} and G_{isr} denote the random variables representing the number of goals scored by two competing teams in match i.

Further, let now $\tilde{y}_i \in \{1 = win, 2 = draw, 3 = loss\}$ be the true ordinal match outcomes from the perspective of the first-named team for match i. Based on the predictions $\hat{\lambda}_{irs}$ and $\hat{\lambda}_{isr}$ from our Poisson model, we can easily calculate the corresponding predicted probabilities $\hat{\pi}_{1i}, \hat{\pi}_{2i}, \hat{\pi}_{3i}$ for these match outcomes via $\hat{\pi}_{1i} = P(G_{irs} > G_{isr})$, $\hat{\pi}_{2i} = P(G_{irs} = G_{isr})$ and $\hat{\pi}_{3i} = P(G_{irs} < G_{isr})$. For this purpose, the so-called Skellam distribution is particularly useful. It is the discrete probability distribution of the integer random variable that is defined as the difference $K_{rs} := G_{rs} - G_{sr}$ of the two (independent) Poisson distributed random variables G_r, G_s (skipping match index i for notational convenience for the moment). Then, the three probabilities $P(G_{rs} > G_{sr}), P(G_{rs} = G_{sr})$ and $P(G_{rs} < G_{sr})$ can be easily obtained by computing $P(K_{rs} > 0), P(K_{rs} = 0)$ and $P(K_{rs} < 0)$ via the Skellam distribution (for more details, see Skellam, 1946).

5.4.2 Ordinal Regression

Inclusion of Covariates

Similar to the Poisson models (5.7) and (5.8), also the basic ordinal model from equation (5.4) can be extended by covariate effects. We then obtain

$$P(Z_{irs} \leq k) = F(\delta + \theta_k + \gamma_{ir} - \gamma_{is}), \quad k = 1, 2, 3, \tag{5.11}$$

with abilities $\gamma_{ir} = \boldsymbol{x}_{ir}^T \boldsymbol{\beta}$ and $\gamma_{is} = \boldsymbol{x}_{is}^T \boldsymbol{\beta}$. Again, the vector $\boldsymbol{\beta} = (\beta_1, \ldots, \beta_p)^T$ collects the corresponding p regression effects of the covariates. As the regression coefficients are currently assumed to be equal for both teams, the linear predictor from model (5.11) can be written for class k in the more compact form $\delta + \theta_k + (\boldsymbol{x}_{ir} - \boldsymbol{x}_{is})^T \boldsymbol{\beta}$. Further extensions are possible, for example, additionally to covariate effects, team-specific ability parameters β_{r0}, β_{s0} can be included, yielding

$$P(Z_{irs} \leq k) = F(\delta + \theta_k + \beta_{r0} - \beta_{s0} + (\boldsymbol{x}_{ir}^T - \boldsymbol{x}_{is}^T)\boldsymbol{\beta}), \quad k = 1, 2, 3. \tag{5.12}$$

A variant of this model has been used in Schauberger et al. (2018).

Prediction of Ordinal Match Outcomes

While for the Poisson models the predicted probabilities $\hat{\pi}_{1i}, \hat{\pi}_{2i}, \hat{\pi}_{3i}$ for the three match outcomes *win*, *draw* and *loss* of a future match need to be derived, e.g., from the Skellam distribution, for ordinal response models these are available more directly. For example, if model (5.12) shall be used for the prediction of a new match, the corresponding match conditions (i.e., the covariates $\boldsymbol{x}_{ir}, \boldsymbol{x}_{is}$) need to be determined and then the corresponding estimated model parameters are plugged into (5.12). Then, exemplarily, the probability for a draw can be derived by the difference

$$P(\text{``Draw''}) = P(Z_{irs} = 2) = P(Z_{irs} \leq 2) - P(Z_{irs} \leq 1).$$

5.4. PREDICTION MODELS

One major disadvantage of the ordinal response models is that it is not possible to receive predictions / probabilities for the exact scores.

5.4.3 Prediction Measures

Based on the predicted probabilities, which are either derived as sketched in Section 5.4.1 for the Poisson models or are directly available for the ordinal response models, several performance measures can be calculated to compare the predictive power of different modelling approaches. Among the most common ones are:

i) the multinomial *likelihood*, which for a single match outcome is defined as $\hat{\pi}_{1i}^{\delta_{1\tilde{y}_i}} \hat{\pi}_{2i}^{\delta_{2\tilde{y}_i}} \hat{\pi}_{3i}^{\delta_{3\tilde{y}_i}}$, with $\delta_{r\tilde{y}_i}$ denoting Kronecker's delta. The multinomial likelihood reflects the probability of a correct prediction. Hence, a large value reflects a good fit.

ii) the *classification rate*, based on the indicator functions $\mathbb{1}(\tilde{y}_i = \arg\max_{r \in \{1,2,3\}} (\hat{\pi}_{ri}))$, indicating whether match i was correctly classified. Again, a large value of the classification rate reflects a good fit. However, note that along the lines of Gneiting and Raftery (2007) the classification rate does not constitute a proper scoring rule.

iii) the *rank probability score* (RPS), which, in contrast to both measures introduced above, explicitly accounts for the ordinal structure of the response $\tilde{y}_i \in \{1, 2, 3\}$. For our purpose it can be defined as $\frac{1}{3-1} \sum_{r=1}^{3-1} \left(\sum_{l=1}^{r} (\hat{\pi}_{li} - \delta_{l\tilde{y}_i}) \right)^2$. As the RPS is an error measure, here a low value represents a good fit.

In this regard it should be pointed out that betting odds provided by bookmakers' companies can be consulted, as they serve as a natural benchmark for the predictive performance of statistical models. For this purpose, one usually would collect so-called "three-way" odds, which consider only the match tendency with possible results *victory team 1*, *draw*, or *defeat team 1* and are usually fixed some days before the corresponding match takes place. This allows the bookmakers to incorporate current information (e.g., injuries of important players) into their odds. Then, by taking the three quantities $\tilde{\pi}_{ri} = 1/\text{odds}_{ri}, r \in \{1,2,3\}$ of a match i and by normalising with $c_i := \sum_{r=1}^{3} \tilde{\pi}_{ri}$ in order to adjust for the bookmaker's margins, the odds can be directly transformed into probabilities using $\hat{\pi}_{ri} = \tilde{\pi}_{ri}/c_i$. Note that these transformed probabilities implicitly assume that the bookmaker's margins are equally distributed on the three possible match tendencies. Finally, these transformed probabilities could be regarded as the bookmakers' match predictions and could be compared to the predictions of statistical models via the performance measures i)-iii) from above.

5.4.4 Application to the German Bundesliga

In the following we will apply one of the Poisson models introduced from above together with a comparable ordinal response model to data from the German Bundesliga. The data set contains information of 1224 matches from the Bundesliga seasons 2014/15–2017/18 and has been constructed by Philip Buczak, a statistics student at TU Dortmund University, during his bachelor thesis. A single match is represented by two rows in the data set, one for the goals and the covariate information of the home team and one for those of the away team. Besides the number of goals, the name of the team, the match day, the season and the match venue, the data set contains the following covariates (among others):

- *Odds*: For every match the average winning odds of both teams were collected from the betting portal BetBrain.com, which were available on the website http://www.football-data.co.uk/germanym.php.

- *Elo:* The Elo club rating is based on the Elo rating system, as discussed in Section 5.3.4, and aims at reflecting the current strength of a football team relative to its competitors. The Elo club ratings are updated after each match day and were available on the website http://clubelo.com/.

- *Market value*: For every match the cumulated market value of the starting line-up of both teams was derived from the website https://www.transfermarkt.de/.

All of these covariates are incorporated in the form of differences, see, e.g., the models (5.8) and (5.12). As an illustration Table 5.3 shows the results and the covariates of the respective teams exemplarily for the first two matches of the Bundesliga season 2014/15.

In the first match of the season, Bayern Munich (BAY) played at home against VFL Wolfsburg (WOB). The average odds for a victory of Munich were 1.25 and for a victory of Wolfsburg 10.87. Hence, from the perspective of the first-named team, Bayern Munich, the difference results in 1.25 - 10.87 = -9.62, while for Wolfsburg the difference yields 9.62. The values of the other covariate differences can be obtained analogously.

We now compare the predictive power of the models (5.8) and (5.12) with respect to the data set from Table 5.3. In addition to

5.4. PREDICTION MODELS

Table 5.3: Exemplary table illustrating the data structure.

Goals	Team	Match day	Season	Odds-diff.	Elo-diff.	Market value-diff.
2	BAY	1	14/15	-9.62	271	199.50
1	WOB	1	14/15	9.62	-271	-199.50
2	HOF	1	14/15	-1.83	-14	56.25
0	AUG	1	14/15	1.83	14	-56.25
⋮	⋮	⋮	⋮	⋮	⋮	⋮

the three covariates *Odds-diff.*, *Elo-diff.* and *Market value-diff.*, we also include team-specific abilities $\beta_{r0}, r = 1, \ldots, N$ for all N teams involved. For each of the four seasons, we predict all matches from the second half of the season by fitting both models to a training data set. For each single matchday a different training data set is created containing all matchdays of the respective season previous to the one at hand. For example, to predict the matches of matchday 18 in the season 2014/15, we learn our models based on all matches from the first 17 matchdays of this season. For the prediction of matchday 19, the learning data set is extended by the matches from matchday 18 and so on. We obtain predictions of a total of $4 \cdot 9 \cdot 17 = 612$ matches. For comparison and as a benchmark, we also regard the corresponding predictions based on the bookmakers' odds.

Exemplarily, Figure 5.1 shows the predictive multinomial likelihood introduced in Section 5.4.3 for the bookmakers' odds as well as for the two regression models (5.8) and (5.12) on the matches from the 17 match days from the second half of the season 2017/18, which served as external validation data. Obviously, the bookmakers' odds provide the best prediction here (median value 0.377), though the two regression approaches come close. The Poisson regression approach (blue box, on the right, median value 0.359) performs slightly better than the ordinal regression model (green box, in the middle, median value 0.353). Hence, at least for the data at hand, modelling the precise number of goals seems to exploit a little bit more of useful information, compared to simply regarding the ordinal outcomes *win*, *draw*, and *loss*.

Moreover, using the (average) betting three-way odds of the 17 validation match days (obtained from *bet365* via http://www.footbaIl-data.co.uk/germanym.php) as well as the corresponding predicted probabilities from our respective models, different betting strategies can be applied (see, e.g., Groll et al., 2018). For every match i and each of the possible three match outcomes $l \in \{1, 2, 3\}$, we can calculate the expected returns as follows: $E[return_{il}] = \hat{\pi}_{il} * odds_{il} - 1$. Then, we choose the outcome with the highest expected return, but only place the bet if the expected maximum return is positive, i.e. if

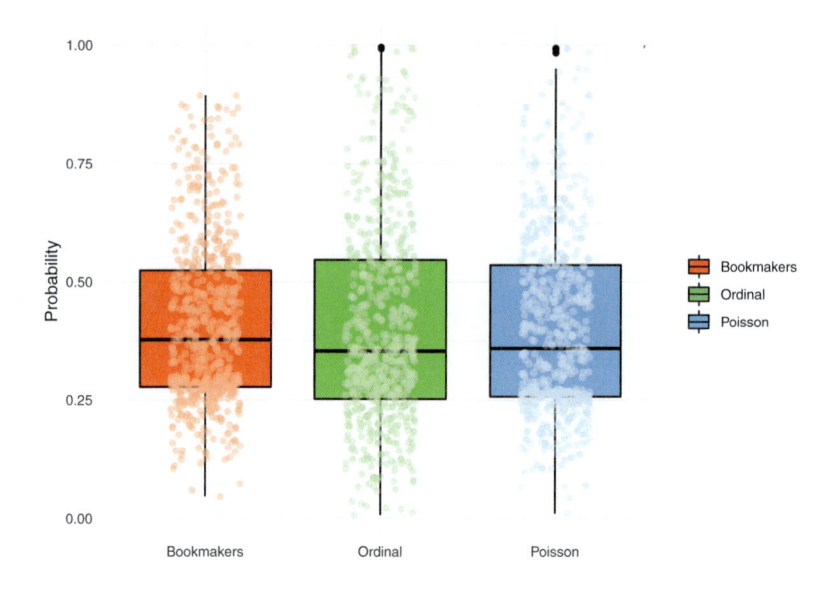

Figure 5.1: Boxplot of the *multinomial likelihood* for the predictions based on the bookmakers' odds (left), the Poisson model (5.8) (middle), and the ordinal response model (5.12) (right).

$\max_{l \in \{1,2,3\}} E[return_{il}] > \tau$ holds, with $\tau = 0$. Koopman and Lit (2015) used different values of the threshold $\tau > 0$ and showed that this way the overall average returns could be increased. While they used constant stake sizes (one arbitrary unit) for each bet, alternative betting strategies with varying stake sizes based on Kelly's criterion (Kelly, 1956) can be applied (see, for example, Boshnakov et al., 2017). With this criterion the optimal stake for single bets can be determined in order to maximise the return by considering the size of the odds and the winning probability.

Figure 5.2 shows the corresponding results. For the ordinal response model (left graphs), both the constant and the Kelly criterion-based strategy yield certain losses for all values of the threshold τ. A rather different result is obtained for the Poisson model. Here, the constant stake strategy outperforms the Kelly criterion-based strategy and for some values of τ even leads to positive returns. For comparison we have also included the average bookmaker margin (red line), which is simply obtained by betting one unit on all of the three outcomes for each match and then averaging the resulting losses. In general,

5.4. PREDICTION MODELS

the return curves all turn out to be rather wiggly, which might indicate that the results should be treated with some caution. Due to the rather small sample size, the results depend quite strongly on single match results and are probably highly variable.

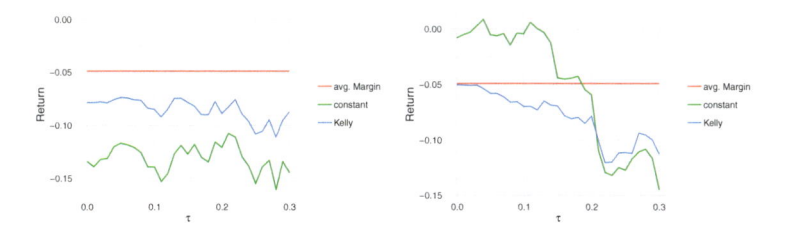

Figure 5.2: Average returns of the ordinal response model (5.12) (left) and the Poisson model (5.8) (right) for constant (green lines) and Kelly-based (turquoise lines) stake sizes for varying threshold value τ; the average bookmaker margin is displayed by the red line.

5.4.5 Extensions

For the basic models from above several extensions exist. In the following we will present some of the most popular extensions, such as the incorporation of regularisation techniques, methods from machine learning, and models that are purely based on bookmakers' odds.

Regularised Modelling

Both the Poisson and the ordinal regression models from Sections 5.2.1 and 5.2.2 can be extended in various ways. When a large number of covariates is supposed to be incorporated into a model and/or if the predictive power of the single variables is not clear in advance, it can be sensible to estimate these models with regularised estimation approaches. For example, Groll and Abedieh (2013), Groll et al. (2015), or Schauberger et al. (2018) use L_1-penalised approaches, while Groll et al. (2018) apply so-called boosting techniques, which stem from the machine learning community. Furthermore, in order to simplify interpretability and reduce complexity of the models, regularisation approaches can be employed to cluster teams with very similar effects with respect to certain covariates. Approaches of that kind are, for example, proposed by Tutz and Schauberger (2015) and Schauberger et al. (2018) for Bradley-Terry-type models with applications to data from the German Bundesliga. If prediction is the major purpose, approaches based on random forests are very powerful, see next section.

Approaches from Machine Learning

Another community that has recently paid increasing attention to the prediction of football events comes from the field of machine learning. This research field combines computer science techniques with statistical methods in order to automatically "learn" from given data. These methods are often very powerful in terms of prediction, but are commonly much harder to interpret than parametric regression models as described above in Sections 5.2.1 and 5.2.2. Therefore, machine learning models are often referred to as black box models, where it is rather difficult to understand which connection the method found between the predictor variables and the outcome. Nevertheless, if prediction is the main purpose, machine learning methods are definitely powerful tools. The possibilities to apply machine learning for the prediction of football matches are as diverse and manifold as the field of machine learning itself. Possible starting points are, for example, the recent publications by Constantinou and Fenton (2017) or Hubáček et al. (2019). The latter evolved from a machine learning competition (*Soccer Prediction Challenge*) launched by the *Machine Learning Journal*, which was followed by a corresponding special issue soliciting papers about machine learning approaches for all aspects of football (Berrar et al., 2019). Schauberger and Groll (2018) show how different types of random forests can be used to predict matches in international football tournaments. The best-performing random forest was then employed in Groll et al. (2019a) to model the FIFA World Cup 2018 in Russia. With a similar hybrid random forest model, Groll et al. (2019b) provided forecasts of the FIFA Women's World Cup 2019 in France.

Purely Bookmaker-Based Modelling

A different type of prediction approaches, which mainly has been applied to major international tournaments like European championships (EUROs) or FIFA World Cups, is solely based on the prospective information contained in bookmakers' odds. For several years, bookmakers have been offering a bet on the winner in advance of the tournament for such major tournaments. Hence, for all participating teams, winning odds are available from several bookmakers' companies.

In Leitner et al. (2010a) this information has first been used for a prediction method. They aggregated the winning odds from several online bookmakers and transformed them into winning probabilities for each team. Based on these winning probabilities, they can compute team-specific abilities by inverse tournament simulation. In inverse tournament simulation, team-specific ability parameters are numerically optimised such that they lead to the given winning probabilities, considering the tournament draw. Therefore, using this technique, the effects of the tournament draw are stripped. With the team-specific abilities all single matches are simulated via simple paired comparisons, and, hence, the complete tournament course is obtained. This strategy has been used to predict the Euro's 2008-2016 (Leitner et al., 2010a; Zeileis et al., 2012, 2016) as well as the FIFA World Cups 2010-2018 (Leitner et al., 2010b; Zeileis et al., 2014, 2018).

Implicitly, this method assumes that all available information on the teams' playing abilities is covered by the bookmakers' expertise. This is not unrealistic, as the bookmakers have strong economic incentives to use sophisticated models when setting up their odds. Consequently, this prediction strategy will always assign the highest probability to become the tournament winner to the team with the lowest (average) bookmaker odds. Note that the corresponding team ability estimates themselves also serve as an informative covariate. They have been included in the hybrid random forest model from Groll et al. (2019b) on the FIFA Women's World Cup 2019 and turned out to be a relevant predictor.

5.5 Conclusion

In this chapter we provided an overview of the most common methods that are used to rank football teams and to predict single matches in association football. For both tasks the basis are well-known fundamental probabilistic approaches that are suitable to model single football matches. Depending on whether the goals scored by the competing teams or the final match outcomes (e.g., in the form *win*, *draw*, *loss*) are regarded, typically either Poisson models or models designed for ordinal response are used. In their most generic form, they typically include team-specific ability parameters and a potential home effect.

Next, the four most popular ranking-methods have been introduced, namely *point-winning systems*, *least squares methods*, *maximum likelihood-based methods*, and *Elo ratings*. The latter three methods involve the calculation of strength parameters for all regarded teams, which allows for a comparison and, hence, a ranking of the teams. Using data from both the German Bundesliga and national team competitions, it was illustrated how the methods can be tailored to account for the specific characteristic of different competition types.

In football another aspect of major importance is an adequate prediction of future matches (or even whole tournaments or leagues). For this purpose, typically, some sort of covariate information is included into the models. For both the Poisson and ordinal models, the standard approaches were presented. We also sketched how future match outcomes could either be predicted or simulated. Again, the methods were illustrated in an application to German Bundesliga data. Finally, some extensions and further developments such as *regularised models*, *machine learning approaches*, or *purely bookmaker-based approaches* were outlined.

Acknowledgement

The authors wish to thank Philip Buczak for providing the German Bundesliga data.

Bibliography

Agresti, A. (1992). Analysis of ordinal paired comparison data. *Applied Statistics*, 41(2):287–297.

Albert, J., Bennett, J., and Cochran, J. J. (2005). *Anthology of Statistics in Sports*. SIAM, Philadelphia.

Bates, D., Mächler, M., Bolker, B., and Walker, S. (2015). Fitting linear mixed-effects models using lme4. *Journal of Statistical Software*, 67(1):1–48.

Bates, D., Maechler, M., Bolker, B., and Walker, S. (2017). *lme4: Linear mixed-effects models using S4 classes*. R package version 1.1-15.

Berrar, D., Lopes, P., Davis, J., and Dubitzky, W. (2019). Guest editorial: special issue on machine learning for soccer. *Machine Learning*, 108(1):1–7.

Boshnakov, G., Kharrat, T., and McHale, I. G. (2017). A bivariate Weibull count model for forecasting association football scores. *International Journal of Forecasting*, 33(2):458–466.

Bradley, R. A. and Terry, M. E. (1952). Rank analysis of incomplete block designs: I. The method of paired comparisons. *Biometrika*, 39(3/4):324–345.

Constantinou, A. and Fenton, N. (2017). Towards Smart-Data: Improving predictive accuracy in long-term football team performance. *Knowledge-Based Systems*, 124:93–104.

Elo, A. E. (1978). *The Rating of Chessplayers, Past and Present*. Arco Publications, New York.

Fahrmeir, L. and Tutz, G. (1994). Dynamic stochastic models for time-dependent ordered paired comparison systems. *Journal of the American Statistical Association*, 89:1438–1449.

Friedman, J., Hastie, T., and Tibshirani, R. (2010). Regularization paths for generalized linear models via coordinate descent. *Journal of Statistical Software*, 33(1):1.

Gneiting, T. and Raftery, A. E. (2007). Strictly proper scoring rules, prediction, and estimation. *Journal of the American Statistical Association*, 102(477):359–378.

Groll, A. and Abedieh, J. (2013). Spain retains its title and sets a new record - generalized linear mixed models on European football championships. *Journal of Quantitative Analysis in Sports*, 9(1):51–66.

Groll, A., Kneib, T., Mayr, A., and Schauberger, G. (2018). On the dependency of soccer scores–A sparse bivariate Poisson model for the UEFA European Football Championship 2016. *Journal of Quantitative Analysis in Sports*, 14(2):65–79.

Groll, A., Ley, C., Schauberger, G., and Van Eetvelde, H. (2019a). A hybrid random forest to predict soccer matches in international tournaments. *Journal of Quantitative Analysis in Sports*, 15(4):271–287.

Groll, A., Ley, C., Schauberger, G., Van Eetvelde, H., and Zeileis, A. (2019b). Hybrid Machine Learning Forecasts for the FIFA Women's World Cup 2019. *arXiv preprint arXiv:1906.01131*.

Groll, A., Schauberger, G., and Tutz, G. (2015). Prediction of major international soccer tournaments based on team-specific regularized Poisson regression: an application to the FIFA World Cup 2014. *Journal of Quantitative Analysis in Sports*, 11(2):97–115.

Hothorn, T., Buehlmann, P., Kneib, T., Schmid, M., and Hofner, B. (2017). mboost: *Model-Based Boosting*. R package version 2.8-1.

Hubáček, O., Šourek, G., and Železný, F. (2019). Learning to predict soccer results from relational data with gradient boosted trees. *Machine Learning*, 108(1):29–47.

Hvattum, L. M. and Arntzen, H. (2010). Using Elo ratings for match result prediction in association football. *International Journal of forecasting*, 26(3):460–470.

Karlis, D. and Ntzoufras, I. (2003). Analysis of sports data by using bivariate Poisson models. *The Statistician*, 52:381–393.

Kelly, J. L. (1956). A new interpretation of information rate. *Bell System Technical Journal*, 35(4):917–926.

Knorr-Held, L. (2000). Dynamic Rating of Sports Teams. *Journal of the Royal Statistical Society: Series D (The Statistician)*, 49(2):261–276.

Koning, R. H. (2000). Balance in competition in Dutch soccer. *Journal of the Royal Statistical Society: Series D (The Statistician)*, 49(3):419–431.

Koopman, S. J. and Lit, R. (2015). A dynamic bivariate Poisson model for analysing and forecasting match results in the English Premier League. *Journal of the Royal Statistical Society: Series A (Statistics in Society)*, 178(1):167–186.

Lasek, J., Szlávik, Z., and Bhulai, S. (2013). The predictive power of ranking systems in association football. *International Journal of Applied Pattern Recognition*, 1(1):27–46.

Leitner, C., Zeileis, A., and Hornik, K. (2010a). Forecasting sports tournaments by ratings of (prob)abilities: A comparison for the EURO 2008. *International Journal of Forecasting*, 26(3):471–481.

Leitner, C., Zeileis, A., and Hornik, K. (2010b). Forecasting the winner of the FIFA World Cup 2010. Research Report Series Report 100, Department of Statistics and Mathematics, University of Vienna.

Ley, C., Van de Wiele, T., and Van Eetvelde, H. (2019). Ranking soccer teams on the basis of their current strength: A comparison of maximum likelihood approaches. *Statistical Modelling*, 19(1):55–77.

McHale, I. and Scarf, P. (2007). Modelling soccer matches using bivariate discrete distributions with general dependence structure. *Statistica Neerlandica*, 61(4):432–445.

McHale, I. G. and Scarf, P. A. (2011). Modelling the dependence of goals scored by opposing teams in international soccer matches. *Statistical Modelling*, 41(3):219–236.

R Development Core Team (2019). *R: A Language and Environment for Statistical Computing*. R Foundation for Statistical Computing, Vienna, Austria. ISBN 3-900051-07-0.

Schauberger, G. and Groll, A. (2018). Predicting matches in international football tournaments with random forests. *Statistical Modelling*, 18(5–6):1–23.

Schauberger, G., Groll, A., and Tutz, G. (2018). Analysis of the importance of on-field covariates in the German Bundesliga. *Journal of Applied Statistics*, 45(9):1561–1578.

Skellam, J. G. (1946). The frequency distribution of the difference between two Poisson variates belonging to different populations. *Journal of the Royal Statistical Society: Series A (General)*, 109(3):296–296.

Stefani, R. T. (1977). Football and basketball predictions using least squares. *IEEE Transactions on Systems, Man, and Cybernetics*, 7(2):117–21.

Tibshirani, R. (1996). Regression shrinkage and selection via the lasso. *Journal of the Royal Statistical Society: Series B (Methodological)*, 58(1):267–288.

Tutz, G. (1986). Bradley-Terry-Luce Models with an Ordered Response. *Journal of Mathematical Psychology*, 30:306–316.

Tutz, G. and Schauberger, G. (2015). Extended ordered paired comparison models with application to football data from German Bundesliga. *AStA Advances in Statistical Analysis*, 99(2):209–227.

van der Wurp, H., Groll, A., Kneib, T., Marra, G., and Radice, R. (2020). Generalised joint regression for count data - a penalty extension for competitive settings. *Statistics and Computing*. To appear.

Zeileis, A., Leitner, C., and Hornik, K. (2012). History repeating: Spain beats Germany in the EURO 2012 Final. Working paper, Faculty of Economics and Statistics, University of Innsbruck.

Zeileis, A., Leitner, C., and Hornik, K. (2014). Home Victory for Brazil in the 2014 FIFA World Cup. Working paper, Faculty of Economics and Statistics, University of Innsbruck.

Zeileis, A., Leitner, C., and Hornik, K. (2016). Predictive Bookmaker Consensus Model for the UEFA Euro 2016. Working Papers 2016-15, Faculty of Economics and Statistics, University of Innsbruck.

Zeileis, A., Leitner, C., and Hornik, K. (2018). Probabilistic forecasts for the 2018 FIFA World Cup based on the bookmaker consensus model. Technical report, Working Papers in Economics and Statistics.

Chapter 6

The Relationship between Running Shoes and Running Injuries: Choosing between a Complicated Truth and a Simple Lie

DANIEL THEISEN
ALAN–MALADIES RARES LUXEMBOURG

RASMUS NIELSEN
AARHUS UNIVERSITY

LAURENT MALISOUX
LUXEMBOURG INSTITUTE OF HEALTH

Abstract

It is a popular belief that the relationship between running footwear and running injuries is direct and causal. This idea is wrong and has been nurtured by the running shoe industry ever since the appearance of the modern running shoe. Still, the latter may well influence the relationship between training load and injury occurrence, thus acting as an effect-measure modifier of running training patterns on injury rather than a direct cause of injury itself. Some recent, well-designed intervention studies using injury as their main outcome variable have provided preliminary results that suggest a relationship between specific shoe features and injury risk. However, these findings mainly apply to certain subgroups of runners, and some stem from secondary analyses that still need confirmation. Shoe prescription guidelines are not scientifically justified in the face of the currently limited evidence base, so caution is warranted against simplistic explanations and erroneous "common sense".

6.1 Introduction

Leisure-time running has numerous positive health benefits (Oja et al., 2015) and is one of the most attractive sports in the world (Hulteen et al., 2017). Among the main reasons for its popularity is the fact that it is very time-efficient, can be performed virtually anywhere, and does not require much specific equipment, except, maybe, for a pair of running shoes.

Many self-declared specialists will argue that it is important to pick "the right" running shoe to avoid the (unfortunately) frequent running injuries. Indeed, within a year's time, roughly one in two runners will experience a running-related injury, although there are large variations across studies and runners' populations (Kluitenberg et al., 2015a). Most injuries result from (chronic) overuse or overload, i.e., a mismatch between the type and amount of training performed and the body's ability to adapt to that physiologic stimulus (Bertelsen et al., 2017). The running shoe being at the interface between the runner and the environment, popular belief has it that the relationship between footwear and running injuries is straightforward and causal. However, a second look reveals that this cannot be the case because a running injury can only occur if a person does running training. In other words, training characteristics are causally related to running injuries, and the footwear only influences how much running a person can tolerate before injury occurs (Malisoux et al., 2016a). Although much more than a subtlety, this aspect has been somewhat overlooked by the scientific community from the field of sports injury prevention and will be addressed in the first part of this chapter along with other important facets of study design and statistical approaches.

Although footwear has its importance in the context of running injury prevention, its role is often overemphasised by sports gear manufacturers. This may be a simple selling strategy driven by the overwhelming market the running shoe industry represents. On the other hand, the science around this topic has generated some degree of confusion, given that 1) there are very few randomised trials to provide high-level evidence on the impact of footwear on injury occurrence (Napier and Willy, 2018), and 2) many conclusions from the literature are based on cross-sectional studies on the influence of footwear technology on running mechanics (Theisen et al., 2014). Undue inference to injury risk has been made by some of the most prominent researchers, given that the outcome of interest was not injury in those studies. The second part of this chapter will thus review the most recent findings from running injury epidemiology to illustrate the influence of running shoes on injury risk.

6.2 Part I: A Look Behind the Curtains of Scientific Methodology

6.2.1 It is all in the Study Design

There is a plethora of scientific studies focusing on the effect of footwear on running mechanics and injury. The non-specialist reader gets easily confused when facing the abundance of literature results, which are not always consistent and occasionally subject to overinterpretation. It is important to scrutinise the publications bearing in mind the participants' profile and the study design (cross-sectional[1] *versus* longitudinal[2]), which will determine the generalisability of the findings and level of evidence of the results as well as the announced outcomes of interest, which will direct the scope of the conclusions that are supported by those results. To facilitate understanding, the "Bermuda Triangle framework" has been proposed as a reading grid (Theisen et al., 2014) (with reference to a vaguely defined area in the North Atlantic Ocean), depicting the types of studies conducted within the "vaguely defined" triangular relationship between running footwear, running biomechanics, and running injury (Figure 1).

The most extensively studied facet has been the influence of running shoes on running biomechanics (Chambon et al., 2014; Malisoux et al., 2017; Mann et al., 2015a; Squadrone et al., 2015). Typically, the study aim is to investigate how certain shoe features influence selected variables related to external ground reaction forces (kinetics) and body motion (kinematics) during running. These biomechanics studies are extremely valuable since they provide direct insight into the acute adjustments of runners to specific footwear technology under different conditions. Generally, they are of cross-sectional design (rare exceptions exist (Malisoux et al., 2017)) and require a small number of (mostly uninjured) participants. They are almost always confined to the biomechanics laboratory, involving high-tech measuring equip-

---

[1] A cross-sectional study (or transverse study) is a type of observational study (i.e., does not involve manipulating variables) that analyses data from a population, or a representative subset, at a specific point in time. They differ from case-control studies in that they aim to provide data on the entire population under study, whereas case-control studies typically include only individuals with a specific characteristic and compare them with a sample, often a tiny minority, of the rest of the population.

[2] A longitudinal study is a research design that involves continuous or repeated monitoring of risk factors or health outcomes, or both. They can be retrospective or prospective.

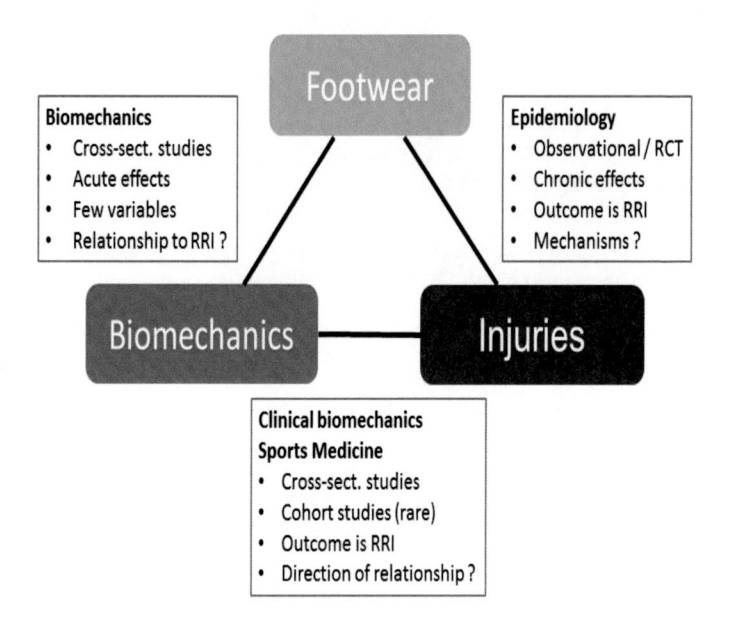

Figure 6.1: The "Bermuda Triangle framework" as a reading grid for studies focusing on running footwear, running mechanics, and running-related injury (RRI). RCT: Randomised controlled trial. (Reprinted from Sport–Orthopadie–Sport Traumatologie–Sports Orthopaedics and Traumatology, 32(2), Daniel Theisen, Laurent Malisoux, Paul Gette, Christian Nührenbörger, Axel Urhausen, "Footwear and running-related injuries–Running on faith?", 169-176, Copyright (2016), with permission from Elsevier.)

ment (force plates and 3-D high-speed motion capture systems), and rely on the implicit assumption that the recorded running technique is representative of the runners' usual running style (ecological validity). Importantly, this kind of studies does not allow to draw conclusions about running injury risk, since this is not the main outcome of interest–any conclusion of that kind is unjustified and speculative.

6.2. PART I

Studies from the field of clinical biomechanics and sports medicine often analyse the relationship between running mechanics and injury by comparing recently injured runners to a healthy control group within a case-control design [3]. The methodologies used are similar to the previous study type, with designs that can be cross-sectional (Mann et al., 2015b), retrospective (Daoud et al., 2012), or prospective (Davis, 2014). In general, attempts are made to match both groups regarding personal characteristics that are known or thought to be related to injury risk (e.g., age, sex, body mass, training status, running experience, etc.). Therefore, any differences in the biomechanics of running could theoretically be associated with injury risk. However, there may be many unknown or non-measured factors for which the groups studied are not necessarily matched and that may influence the results. Furthermore, the direction of the relationship cannot be determined in cross-sectional and retrospective studies, which strongly limits the scope of the conclusions that can be drawn.

Epidemiological research approaches in the area of running injuries involve much greater numbers of participants (at least several hundreds). These are generally followed over several months to up to a year within an observational study (Kluitenberg et al., 2015b; Malisoux et al., 2015a; Nielsen et al., 2019c) or a randomised trial (Malisoux et al., 2015b; Ryan et al., 2014; Theisen et al., 2016). The latter design has the advantage that the randomisation will allow for an equal distribution between the study groups of all other (known or unknown) factors that may influence injury risk. The main outcome of interest is running injury, which then also permits to draw conclusions about injury risk, for example, in relation to certain personal characteristics, training behaviour, or footgear used. These observational cohort studies and randomised trials bear a higher level of evidence than cross-sectional studies, which makes causal inference generally more plausible, especially for randomised trials. Unfortunately, they do not provide any explanations about the underlying mechanism of risk factors that are found as significant. The results and their interpretations may be influenced by the injury definition used (Kluitenberg et al., 2016) and the study population involved (Videbaek et al., 2015). The background methodology to this type of studies will be further presented in the following paragraphs.

---

[3] A case-control study compares patients with the health outcome of interest (as well as their past exposure to suspected aetiological factors) with that of controls who do not have the health outcome.

6.2.2 Time to Event Analyses are the Current Gold Standard

As already pointed out, running injuries are common, with injury incidence density ranging from 3 to over 30 injuries per 1000 hours of running (Videbaek et al., 2015). It is on average higher in novice runners (18 injuries/1000 hrs) than in more experienced runners (8 injuries/1000 hrs). Acute injuries account for about 15-25%, while overload injuries represent some 75-85% (Malisoux et al., 2020, 2015b). The vast majority of running injuries are related to cumulative microtrauma and concern mainly the lower limbs and back regions (Lopes et al., 2012). They obviously have a complex, multifactorial origin. Most of them are thought to be caused by training errors, although a recent systematic review was unable to identify any trends from the existing literature (Nielsen et al., 2019a).

Commonly reported measures of association in sports injury research are relative risk, risk difference, incidence rate ratio, incidence rate difference, and hazard rate ratio (Nielsen et al., 2019b). Estimators of simple relative risk and risk difference and their 95% confidence intervals are quite easy to compute and interpret, but these measures are helpful only if sport exposure is similar in the groups compared. Incidence rate ratio and incidence rate difference address this shortcoming by accounting for sport exposure, which can be expressed, for example, as 1000 hours of running practice. However, these statistics do not consider the case of participants dropping out of the study during the follow-up (censoring) (Jungmalm et al., 2020), nor do they permit to account for different explanatory variables simultaneously. Therefore, more advanced statistical approaches such as time to event models are currently the preferred methods to analyse different categorical and continuous variables in relation to running injury risk (Nielsen et al., 2016).

The Cox regression model[4] based on survival analysis ("survival" until the occurrence of the event of interest, i.e., injury) with its hazard rate ratio is commonly used in prospective cohort studies and trials in sports injury research, including running injury. The hazard rate ratio has the advantage of providing a single estimate (plus its confidence interval) to compare groups or conditions regarding injury rate, provided that hazard rates are proportional, which can be verified by

---

[4]Cox regression (or Cox proportional hazard) models are a class of survival models. They relate the time that passes before some event occurs (e.g., injury) to one or more covariates (e.g., footwear) that may be associated with that quantity of time (e.g., hours of running). The proportional hazards condition states that covariates are multiplicatively related to the hazard.

6.2. PART I

log-minus-log plots of all variables included in the model. Moreover, this approach is superior to binomial regression[5] or logistic regression models[6] because it accounts for censoring, i.e., a situation where a participant drops out of the study before the end of the observation period or the occurrence of the event of interest (Nielsen et al., 2019b). Whereas the classical time scale used with Cox proportional hazards regression models is calendar time, it is more appropriate to use another scale in the context of running injury research, for example, hours spent running.

More recently, the pseudo-observation method has been proposed as an interesting alternative to the Cox regression (Nielsen et al., 2019b). This method allows to compute two different proportion-based estimates: the cumulative relative risk and the cumulative risk difference, representing respectively a ratio and an absolute difference in risk between two groups at one or several time points on the abscissa of the Kaplan-Meier plot[7]. With this model the assumption about proportionality of hazard rates across the whole-time scale is irrelevant, but one or several time points of relevance need to be determined (and preferably justified) beforehand in the published protocol.

The Cox regression model and the pseudo-observation method require that a minimum number of events of interest be included in the analysis to yield meaningful results. Based on simulation studies, guidelines have been provided for the hazard rate ratio and the cumulative risk difference (10 events of interest per regressor) as well as for the cumulative relative risk (15 events of interest per regressor) (Nielsen et al., 2019b). This is an important aspect that needs to be considered already at an early stage of designing a study protocol, especially if stratified analyses[8] are to be run (where the sample is separated into subgroups or strata, according to the confounders or effect-measure modifiers of interest), as the number of variables included in the model increases exponentially for each additional stratum analysed, potentially leading to sparse data bias (Nielsen et al., 2012).

---

[5] Binomial regression is used to assess the relationship between a binary response variable (e.g., sustained an injury or did not sustain an injury) and other explanatory variables (e.g., footwear, running frequency, etc.).

[6] Logistic regression is a binomial model that in its basic form uses a logistic function to model a binary dependent variable.

[7] The Kaplan-Meier plot shows what the probability of an event is at a certain time interval in each study group. If the sample size is large enough, the curve should approach the true survival function for the population under investigation.

[8] Stratified analysis is a powerful statistical approach that allows testing for confounding and interaction.

6.2.3 Running Shoes do not Cause Running Injuries

Much research has been directed towards studying risk factors for running injuries, including sex, age, body mass, anatomy of the lower limbs, lifestyle characteristics, health-related factors, training behaviour, or the type of running shoes used. Only very few factors have been consistently found to be related to injury risk, most notably previous running injury (Hulme et al., 2017; Saragiotto et al., 2014a; van Gent et al., 2007). However, many of these factors are by themselves insufficient to trigger an injury because the causal architecture underpinning running injury occurrence is not considered (Malisoux et al., 2016a). For example, being overweight or wearing a certain shoe type does not *per se* cause running injury. Performing running practice is a necessary condition and, in fact, the only necessary one. Therefore, training load characteristics are more relevant as primary exposures of interest when causal mechanisms are to be explored, bearing in mind that inappropriate training load may lead to cumulated biomechanical strain that exceeds the body's capacity to adapt to the physiological overload (Bertelsen et al., 2017). In short, it is "too much" running training that causes running injuries.

Other factors more distantly related to running injury may influence the load capacity of the organism or the relationship between running participation and injury risk. Although they may appear as statistically significant in a multiple regression model, they are really "only" effect modifiers when considered within an etiological[9] framework of running injury risk (addressing the causal relationship with injury), as represented in Figure 2 below. As an illustrative (although absurd) example, consider running a marathon in running shoes *versus* in alpine ski boots – it is quite clear which condition represents a greater injury risk.

Given the foregoing argument, future research looking into the causes of running injuries need to go beyond the traditional approach of merely identifying risk factors based on (stepwise) multivariable regression analyses. Not all significant variables in a model are confounders for the outcome and directly associated with it. The effect of training-related variables on injury risk may be different across strata of non-training-related variables, such as running shoe type used. Stratified analyses will also allow to explore synergism between two or more variables and to compute the relative or absolute excess risk due to interaction or effect-measure modification (Knol and VanderWeele, 2012; Malisoux et al., 2016a; Nielsen et al., 2019b).

---

[9]Etiology is the study of causes or origins of various phenomena.

```
                    ┌─────────────────────────────────┐
                    │  Running shoe characteristics   │
                    └─────────────────────────────────┘
                         (assumed effect-measure modifiers)
```

┌─────────────────────────────────────┐   ┌─────────────────────────────┐
│ Running training characteristics   │   ⟹   │ Running-related injuries   │
└─────────────────────────────────────┘   └─────────────────────────────┘
  (assumed primary risk factors causally   (Outcome of interest)
       related to running injuries)

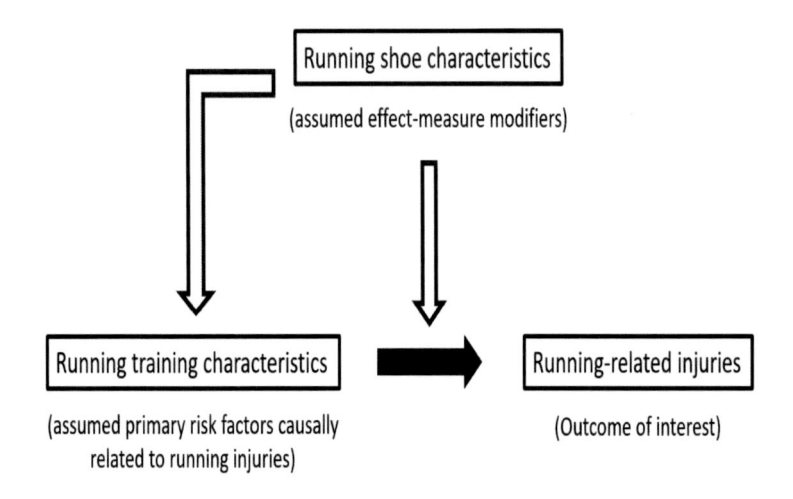

Figure 6.2: The causal relationship between running exposure characteristics (e.g., distance run, running velocity, training frequency, sudden or chronic changes in these variables, etc.) and running injuries; non-training-related characteristics (amongst others running shoe technology) influence the training load that a runner can tolerate before injury occurrence and might also modify training exposure characteristics (based on (Malisoux et al., 2016a)).

6.2.4 Accepting Impermanence: Time-Varying Exposures, Effect Modifiers and Outcomes

As already highlighted above, running shoes with their specific characteristics are supposed to be effect modifiers of the relationship between a runner's training load exposure and running injury. This suggests that stratified analyses as per shoe model would be required to study the relationship between training behaviour and injury, and thus to analyse if certain footwear technology influences the runner's training load tolerance. However, training behaviour is rarely constant (just like many other personal characteristics, such as body mass, fatigue, or nutrition), and many running exposure characteristics change over time as a runner is being followed-up in the framework of a study. This fact calls for advanced statistical approaches, such as time-to-event analyses, where training load levels are included in the model as categorised (or continuous) variables using a delayed entry function, thus enabling the study of non-linear dose-response relationships in the association between training load and injury risk (Nielsen et al., 2016). These levels in training load can be examined either as fixed states (e.g. running volume-based categories) or as *changes or transitions between states* (e.g. change in running volume-based category from one week to another), but these definitions should be specified beforehand along with the study objectives and hypotheses. Non-training-related variables, such as shoe type used, can be included as time-varying effect-measure modifiers in the models. The latter eventually become very sophisticated and rely on complex and only recently proposed methodology (Nielsen et al., 2016). Furthermore, as already pointed out above, a major challenge will be to have sufficient data to allow for inclusion of these variables, which only becomes feasible with ambitious study protocols involving thousands of participants and long follow-up periods(Nielsen et al., 2014).

When studying risk factors for running injury, it is often assumed that the outcome of interest is binary: a runner is either injured or not. To solve some ambiguity around these 2 states, experts have proposed consensus definitions for injury in many different sport disciplines, including running: *"Running related (training or competition) musculoskeletal pain in the lower limbs that causes a restriction on or stoppage of running (distance, speed, duration, or training) for at least 7 days or 3 consecutive scheduled training sessions, or that requires the runner to consult a physician or other health professional"* (Yamato et al., 2015). The main advantage of such a definition is that it creates a common understanding amongst researchers but also between scientists and the study participants who need to report the outcome whenever present. However, reality is much more complex, and runners generally manage to "train around" episodes of chronic pain without necessarily reducing or interrupting their training schedule. During these phases, the runner is not truly injury free but not necessarily injured either according to the definition above. To fully capture the phenomenon, it is necessary to record these "intermediate" states, as previously proposed in other sport contexts (Clarsen et al., 2015). This would mean that the injury status becomes a time-dependent variable, which then also allows to account for multiple injuries (or injury states) over time, recurrent injuries, and injuries not related to running. The latter requires a competing risk analysis[10].

Time-dependent exposures, effect modifiers, and outcomes can be studied using multistate models, where a runner is classified into several exposure and outcome states over time and transitions be-

---

[10] A competing risk is an event that either hinders the observation of the event of interest or modifies the chance that this event occurs. Thus, competing risk analysis refers to a special type of survival analysis that aims to correctly estimate the probability of an event in the presence of competing events.

tween them during the follow-up. Since the possibilities of including training- and non-training-related variables as well as outcome data become virtually endless in a multistate-transitions model, it is necessary to make reasonable methodological choices *a priori* in accordance with the specific study objectives and hypotheses (Nielsen et al., 2019b, 2012, 2016).

6.3 Part II: The Front-Row View

6.3.1 Strong Beliefs and Competition

Most recreational runners are extremely critical towards the kind of shoes they use for their sport. This is *a priori* a good principle that should help them pick a model that is most adapted to their needs. Saragiotto and co-workers investigated the runners' beliefs regarding running-related injuries and found that next to training factors and body limits, recreational runners largely attribute injury risk to running shoes (Saragiotto et al., 2014b). Many consider the crude relationship between shoe characteristics and injury as direct and causal. This assumption is fallacious, as discussed above, and it is likely that the role of running shoes on injury risk has been overrated.

Strong beliefs about the role of running shoes may at least partly stem from the messages and selling arguments put forward by the running shoe industry. The market is huge, considering that running is among the top 3 most popular physical activities around the globe (Hulteen et al., 2017). Competition is fierce, and thus the aim of the different brands is to differentiate themselves from their competitors by promoting specific technology that is supposed to solve specific problems or address allegedly special needs of certain subgroups of runners. The following part will present the current state of knowledge about the relationship between running shoes and injury risk and point out some open questions that still warrant further research.

6.3.2 Running Shoe Prescription does not Work

Most running injuries have multiple causes and result from repetitive overload. This means that there is a mismatch between the body's physical load and regeneration capacity on the one hand and external and internal mechanical strain generated by running training on the other hand. According to the popular shoe prescription approach, previously termed the "shoe-shop theory" (Theisen et al., 2014), essentially based on expert opinion (Davis, 2014; Richards et al., 2009), running injuries are caused by excessive external ground reaction forces and excessive foot motion. Therefore, running shoes should be designed to reduce impact forces and attenuate (excessive) foot pronation during the stance phase. Based on foot morphology, three main shoe types have emerged(Knapik et al., 2014). "Cushioned shoes" have a softer midsole and are advised to runners with high-arched, rigid feet with reduced pronation[11]. "Motion-control shoes" are provided with dual-density midsoles, arch support features, or a rigid heel counter to limit rearfoot eversion. They are recommended to runners with flat feet displaying excessive foot pronation and lower limb malalignment during stance phase. Finally, "stability shoes" have some cushioning and some motion control and are suited for runners with normal foot morphology and running mechanics.

Surprisingly, until recently there was no scientific evidence whatsoever to support the shoe-shop theory (Richards et al., 2009). If this theory is correct, shoe prescription tailored to the runner's foot type should lead to a reduction of running injury risk. This was systematically tested in the context of the US military services, more specifically in Army, Air Force, and Marine Corps basic training (Knapik et al., 2014). Three trials were conducted in 2007 using the same methodology, whereby 7.203 recruits were randomly assigned to an experimental and a control group. Those in the experimental group were provided with specific shoe types based on their plantar shape: depending on whether they had a low, medium, or high foot arch, they received motion-control, stability, or cushioned running shoes. Those in the control group were assigned a stability shoe irrespective of their plantar shape. The results showed that there was no difference between the 2 groups, yielding the overall conclusion that selecting shoes based on plantar shape is ineffective in lowering injury risk. Furthermore, injury rates did not differ across the 3 shoe types used. Thus, it seems that the shoe-shop theory is not defensible in the context of military training. A similar approach was applied in a cohort of female runners, and again the results did not support the theory, although the sample size was limited, and the outcome was pain level (Ryan et al., 2011).

Even if shoe prescription according to the prevailing theory does not work, this does not mean that individual shoe features are irrelevant in the context of injury prevention. It is therefore interesting to consider the effect of different shoe characteristics on injury risk, which will be addressed hereafter.

---

[11] Foot pronation during the stance phase is in fact a natural shock absorption mechanism resulting from the anatomic disposition of the foot and the lower leg.

6.3.3 Is Cushioning Important?

The rationale for promoting cushioned shoes is that they can reduce external impact forces, which are themselves considered a major cause of injury. While the latter idea has received some scientific support (Davis et al., 2016; van der Worp et al., 2016), it has long been known that changes in shoe cushioning (mainly through changes in midsole hardness) have little systematic influence on impact force characteristics (Nigg, 2001; Nigg et al., 1987), as opposed to, e.g., running velocity (Nigg et al., 1987) or step rate (Heiderscheit et al., 2011). In short, running style is much more important. This does not exclude a role of shoe cushioning in injury risk, but the action mechanism is most likely not through external impact forces.

It seems that currently there are only three trials looking at the effect of footwear shock absorption characteristics on injury risk. A study on 1200 Air Force recruits analysed if shock absorbing insoles had an effect on lower limb injuries during basic military training (Withnall et al., 2006). Injury rates were similar across the three study groups, providing no support for the use of shock absorbing insoles. A more recent trial from Luxembourg investigated injury risk in 247 recreational runners randomly allocated to one of two groups: one receiving a standard running shoe with a soft midsole, the other using a shoe with a harder midsole (Theisen et al., 2016). Noteworthy is that the shoe prototypes were specifically designed for the trial and identical in all aspects except for midsole density, providing a 13% difference in impact attenuation properties. Furthermore, the study shoes were anonymised so that both participants and the researchers did not know who had received which shoe type until after the trial[12]. After a five-month observation period, injury risk was found to be similar in the two groups. Reasons for these negative results could be that runners adapt their running technique to keep external impact forces constant (Kong et al., 2009) and thereby mitigate the effect of running shoes with lower shock absorption. Another possibility is that the shoe properties were not sufficiently dissimilar to reveal any effect of shoe cushioning. This particular aspect was addressed in a follow-up study by the same team from Luxembourg (Malisoux et al., 2020), using the exact same methodology, except for a shoe stiffness difference of 35%. The researchers also recruited many more participants, 848 in total, and found that injury rate was higher in those runners having received the hard study shoes after 6 months of follow-up. Furthermore, a stratified analysis according to body mass showed that the effect of greater risk in hard shoes was confined to

---

[12] This is the so-called double-blinding procedure, where participants and assessors are blinded to the treatment to reduce the risk of bias.

light runners (Malisoux et al., 2020). In other words, for the first time, these results suggest a protective effect of shoe cushioning. This is only true for light runners but not for heavy ones. In how far these observations are related to impact force attenuation during running is currently being investigated by the same team.

6.3.4 Foot Morphology, Injury Risk and Shoe Types

Pronated foot posture and excessive foot eversion during stance phase is considered to compromise lower extremity alignment and increase the risk of certain running injuries. Footgear equipped with pronation control technology has been advertised on the basis that it counteracts foot pronation. However, as for shoe cushioning, research from biomechanics has shown that different shoe orthoses to support the medial foot (in particular the foot arch) do not "correct" foot eversion motion and yield only small and inconsistent within-subject effects, much smaller than the observed between-subject differences (Nigg, 2001; Stacoff et al., 2000). Again, this does not mean that pronation control technology is irrelevant to injury risk, but it does not act via attenuating foot pronation.

The question remains if foot posture is indeed a risk factor for running injuries. A recent meta-analysis[13] suggests that there is a relationship between a pronated foot posture and the risk of developing medial tibial stress syndrome in different sports, including running, although the effect size is small (Neal et al., 2014). On the other hand, a large one-year prospective observational study from Denmark (DANORUN-study) on over 900 runners found no link between foot pronation and injury risk in novice runners (Nielsen et al., 2019c).

If pronation control features in running shoes have little influence on foot motion during running, and if foot posture is only weakly related to injury risk, it is legitimate to wonder why so many shoe models come with this technology and if it has an influence at all on injury risk. This question was addressed within a trial where 372 runners were randomly assigned to one of two groups that received identical running shoes except that one shoe model was equipped with motion control technology, while the other was not (Malisoux et al., 2015b). At the end of the six-month observation period, it was found that the group using motion control shoes had a lower injury risk and that the positive effect was confined to those runners with pronated feet. Although the study sample size was too small to draw a definite conclusion, these results are somewhat surprising, considering the foregoing arguments. Still, this is the first study that provides some justification

---

[13] Meta-analysis is the statistical procedure for combining data from multiple studies.

for the use of motion control technology in cushioned running shoes. Unfortunately, the underlying mechanism of reduced injury risk is not known. Additionally, the authors of that study pointed out that their results might not be generalisable, because many standard-cushioned running shoes qualified on the market as "neutral" have some motion control features, whereas the shoes from their control group had none.

6.3.5 Other Shoe Features

Next to shoe cushioning and pronation control features, other aspects related to running footwear have been investigated, but their findings have not been confirmed in subsequent studies so far. The influence of heel-to-toe drop of standard cushioned running shoes was tested in a randomised controlled trial involving 553 leisure-time runners (Malisoux et al., 2016b). There was no difference in injury risk overall when comparing the three groups that have received shoes with a drop of either 10 mm, 6 mm, or 0 mm. However, when stratifying their participants into occasional runners and regular runners, the researchers found that low drop shoes were associated with lower injury risk in the former and with higher injury risk in the latter. Put differently, low-drop shoes could be more hazardous in regular runners and preferable for occasional runners.

The effect of shoe degradation over 200 miles was tested in 24 runners and found to induce modifications in the running pattern, i.e. an increase in stance time, to maintain constant variables related to external impact forces acting on the body (Kong et al., 2009). Furthermore, the adaptations to shoe use were not influenced by different cushioning technologies, which hence disqualifies shoe cushioning technology as a critical aspect in shoe degradation. The influence of shoe status on injury risk has not yet been assessed.

A simple but pragmatic study was focused on shoe cost and analysed if more expensive shoes (£70-75) improve plantar pressure attenuation and perceived comfort compared to low-priced models (£40-45 and £60-65) (Clinghan et al., 2008). Although the results from the 43 participants are somewhat difficult to interpret, the authors concluded that cushioning performance was not related to shoe cost and that the low- and medium-cost shoes provided equally good subjective comfort sensation compared to the high-cost models. Neither shoe cost nor the outcome measures from this study have been investigated in relation to injury risk.

6.3.6 Throw Your Shoes Away?

Some may ask the question if they need any running shoes at all. The concept of "barefoot running" or running in minimalist shoes has gained status over the past 20 years, with influential scientific articles and popular books promoting the general idea that our ancestor hunters and gatherers from over 10,000 years ago obviously excelled in bipedal locomotion and running, a necessary quality for survival. Therefore, running barefoot may represent an optimal solution (compared to running in our current modern running shoes) to lower running injury risk (Lieberman, 2012). From a human evolutionary history perspective, we have been running barefoot or in minimalist footwear virtually forever, and some tribes still do today (Lieberman et al., 2010). It is true that the modern running shoe has only been introduced since the 1960s-70s (Davis, 2014), which might actually have produced a mismatch between the running mechanics with which we evolved and the adaptation to our modern environment (Davis et al., 2017). Despite the apparent logic that "natural" barefoot running or using minimalist footwear may decrease injury risk (Lieberman, 2012), it is surprising how previous arguments from the running shoe industry regarding the importance of shoe technology are suddenly thrown overboard by the running shoe industry. Well, a new market is born, and more and more brands now advertise their models of minimalist shoes with no heel-to-forefoot drop, no arch support, no cushioning midsole, and a flexible heel counter or none at all (Davis, 2014).

The introduction of minimalist shoes and the claims about potentially decreasing injury risk have generated much controversy (Tam et al., 2014). Running barefoot or in minimalist shoes is often (Lieberman et al., 2010; Squadrone and Gallozzi, 2009), but not always (Hatala et al., 2013), associated with adopting a midfoot or forefoot strike simply because landing on the heel is uncomfortable. This results in different movement patterns (e.g., a higher step rate) that may attenuate external ground reaction forces and physical load at the knee joint, but also increase strain in the foot region and put greater load on the ankle extensor complex (Divert et al., 2005; Perkins et al., 2014). Lower external impact forces should logically decrease the physical load on the musculoskeletal system and come along with a lower injury risk. The problem that is often overlooked though is that the redistribution of internal strain between different body parts will actually increase injury risk specifically in those areas that are suddenly overloaded (Ryan et al., 2014). More generally, any radical change in running shoe characteristics will inevitably increase the injury risk, unless the runner foresees a progressive habituation period which is yet to be defined based on scientific evidence (Napier and Willy, 2018). When switching from shod to barefoot running or even from maximalist to minimalist footwear, this period should best be planned over several months rather than weeks to allow for appropriate adaptations of the musculoskeletal system (Azevedo et al., 2016). Another advice is to use several shoe pairs during any period of time to allow for a shoe rotation and to smoothen any effect of introducing a new pair into the shoe pool, a strategy that was found to protect runners from running injury (Malisoux et al., 2015a).

6.4 Common Sense Must Prevail

Despite the seemingly sound logic, many explanations and arguments advanced in favour of certain footwear characteristics are simplistic and not substantiated by scientific evidence. As described above, there are now preliminary results from epidemiological studies that suggest that some shoe characteristics may be of benefit to some subgroups of runners. These studies need to be confirmed by further research before any shoe prescription guidelines are scientifically justified. Furthermore, the underlying mechanisms of these few positive results are yet to be uncovered. Finally, the ultimate question as to how much running training (whatever that means in terms of frequency, volume, running speed, load progression, etc.) can be tolerated without injury occurrence with a given anatomical predisposition, biomechanical running style, and specific footwear remains unanswered.

In the face of such a complicated reality, it is difficult to avoid frustration and eventually make the right choice. One should add that a major problem with such divergent and inconsistent information from the scientific literature is that individual (positive or negative) indirect effects of footwear on injury may well be present but are masked by the overall results. Researchers tend to seek for global (and preferably large) effects, with a difference between tested conditions that they consider as real (not due to chance), i.e., a difference that is "statistically significant". However, every runner is unique and reacts differently to a given shoe type, which does not necessarily translate into significant group effects that would support any strong scientific claim. In that respect science can provide some general guidelines, but the final decision will always be an individual one and should preferably be based on correct and unbiased information. At the end of the day, the most relevant aspects to reduce injury risk may well be to choose the most comfortable shoe (Nigg et al., 2015), a prescription that makes sense to most runners, and to keep listening carefully to what your body tells you, which not all runners will accept easily, especially if it means interrupting their training.

It is important for both researchers and runners alike to understand that running shoes do not have a direct causal effect on running injuries. Their main influence is to modify the training load a runner can tolerate in interaction with personal characteristics. Although some will gladly accept a simple lie regarding this influence, the truth is far more complex, as discussed in this chapter – be wary of simplistic explanations and erroneous "common sense", not all that glitters is gold. So, *caveat emptor*, "let the buyer beware" is the general recommendation. The authors of this contribution hope to have shed some light on these issues to help runners make informed decisions concerning their running practice and footwear to use.

Bibliography

Azevedo, A., Mezencio, B., Amadio, A., and Serrao, J. (2016). 16 weeks of progressive barefoot running training changes impact force and muscle activation in habitual shod runners. *PLoS One*, 11(12):e0167234.

Bertelsen, M., Hulme, A., Petersen, J., Brund, R.K., Sorensen, H., Finch, C., Parner, E., and Nielsen, R. (2017). A framework for the etiology of running-related injuries. *Scandinavian Journal of Medicine & Science in Sports*, 27(11):1170–1180.

Chambon, N., Delattre, N., Gueguen, N., Berton, E., and Rao, G. (2014). Is midsole thickness a key parameter for the running pattern? *Gait &Posture*, 40(1):58–63.

Clarsen, B., Bahr, R., Heymans, M., Engedahl, M., Midtsundstad, G., Rosenlund, L., Thorsen, G., and Myklebust, G. (2015). The prevalence and impact of overuse injuries in five Norwegian sports: Application of a new surveillance method. *Scandinavian Journal of Medicine & Science in Sports*, 25(3):323–330.

Clinghan, R., Arnold, G., Drew, T., Cochrane, L., and Abboud, R. (2008). Do you get value for money when you buy an expensive pair of running shoes? *British Journal of Sports Medicine*, 42(3):189–193.

Daoud, A., Geissler, G., Wang, F., Saretsky, J., Daoud, Y., and Lieberman, D. (2012). Foot strike and injury rates in endurance runners: a retrospective study. *Medicine & Science in Sports & Exercise*, 44(7):1325–1334.

Davis, I. (2014). The re-emergence of the minimal running shoe. *Journal of Orthopaedic & Sports Physical Therapy*, 44(10):775–784.

Davis, I., Bowser, B., and Mullineaux, D. (2016). Greater vertical impact loading in female runners with medically diagnosed injuries: a prospective investigation. *British Journal of Sports Medicine*, 50(14):887–892.

Davis, I., Rice, H., and Wearing, S. (2017). Why forefoot striking in minimal shoes might positively change the course of running injuries. *Journal of Sport and Health Science*, 6(2):154–161.

Divert, C., Mornieux, G., Baur, H., Mayer, F., and Belli, A. (2005). Mechanical comparison of barefoot and shod running. *International Journal of Sports Medicine*, 26(7):593–598.

Hatala, K., Dingwall, H., Wunderlich, R., and Richmond, B. (2013). Variation in foot strike patterns during running among habitually barefoot populations. *PLoS One*, 8(1):e52548.

Heiderscheit, B., Chumanov, E., Michalski, M., Wille, C., and Ryan, M. (2011). Effects of step rate manipulation on joint mechanics during running. *Medicine & Science in Sports & Exercise*, 43(2):296–302.

Hulme, A., Nielsen, R., Timpka, T., Verhagen, E., and Finch, C. (2017). Risk and protective factors for middle- and long-distance running-related injury. *Sports Medicine*, 47(5):869–886.

Hulteen, R., Smith, J., Morgan, P., Barnett, L., Hallal, P., Colyvas, K., and Lubans, D. (2017). Global participation in sport and leisure-time physical activities: A systematic review and meta-analysis. *Preventive Medicine*, 95:14–25.

Jungmalm, J., Bertelsen, M., and Nielsen, R. (2020). What proportion of athletes sustained an injury during a prospective study? Censored observations matter. *British Journal of Sports Medicine*, 54(2):70–71.

Kluitenberg, B., van Middelkoop, M., Diercks, R., and van der Worp, H. (2015a). What are the differences in injury proportions between different populations of runners? A systematic review and meta-analysis. *Sports Medicine*, 45(8):1143–1161.

Kluitenberg, B., van Middelkoop, M., Smits, D., Verhagen, E., Hartgens, F., Diercks, R., and van der Worp, H. (2015b). The nlstart2run study: Incidence and risk factors of running-related injuries in novice runners. *Scandinavian Journal of Medicine & Science in Sports*, 25(5):515–523.

Kluitenberg, B., van Middelkoop, M., Verhagen, E., Hartgens, F., Huisstede, B., Diercks, R., and van der Worp H. (2016). The impact of injury definition on injury surveillance in novice runners. *Journal of Science and Medicine in Sport*, 19(6):470–475.

Knapik, J., Trone, D., Tchandja, J., and Jones, B. (2014). Injury-reduction effectiveness of prescribing running shoes on the basis of foot arch height: summary of military investigations. *Journal of Orthopaedic & Sports Physical Therapy*, 44(10):805–812.

Knol, M. and VanderWeele, T. (2012). Recommendations for presenting analyses of effect modification and interaction. *International Journal of Epidemiology*, 41(2):514–520.

Kong, P., Candelaria, N., and Smith, D. (2009). Running in new and worn shoes: a comparison of three types of cushioning footwear. *British Journal of Sports Medicine*, 43(10):745–749.

Lieberman, D. (2012). What we can learn about running from barefoot running: an evolutionary medical perspective. *Exercise and Sport Sciences Reviews*, 40(2):63–72.

Lieberman, D., Venkadesan, M., Werbel, W., Daoud, A., D'Andrea, S., Davis, I., Mang'eni, R., and Pitsiladis, Y. (2010). Foot strike patterns and collision forces in habitually barefoot versus shod runners. *Nature*, 463(7280):531–535.

Lopes, A., Hespanhol Junior, L., Yeung, S., and Costa, L. (2012). What are the main running-related musculoskeletal injuries? A systematic review. *Sports Medicine*, 42(10):891–905.

Malisoux, L., Chambon, N., Delattre, N., Gueguen, N., Urhausen, A., and Theisen, D. (2016a). Injury risk in runners using standard or motion control shoes: a randomised controlled trial with participant and assessor blinding. *British Journal of Sports Medicine*, 50(8):481–487.

Malisoux, L., Chambon, N., Urhausen, A., and Theisen, D. (2016b). Influence of the heel-to-toe drop of standard cushioned running shoes on injury risk in leisure-time runners: A randomized controlled trial with 6-month follow-up. *American Journal of Sports Medicine*, 44(11):2933–2940.

Malisoux, L., Delattre, N., Urhausen, A., and Theisen, D. (2020). Shoe cushioning influences the running injury risk according to body mass: A randomized controlled trial involving 848 recreational runners. *American Journal of Sports Medicine*, 48(2):473–480.

Malisoux, L., Gette, P., Chambon, N., Urhausen, A., and Theisen, D. (2017). Adaptation of running pattern to the drop of standard cushioned shoes: A randomised controlled trial with a 6-month follow-up. *Journal of Science and Medicine in Sport*, 20(8):734–739.

Malisoux, L., Nielsen, R., Urhausen, A., and Theisen, D. (2015a). A step towards understanding the mechanisms of running-related injuries. *Journal of Science and Medicine in Sport*, 18(5):523–528.

Malisoux, L., Ramesh, J., Mann, R., Seil, R., Urhausen, A., and Theisen, D. (2015b). Can parallel use of different running shoes decrease running-related injury risk? *Scandinavian Journal of Medicine & Science in Sports*, 25(1):110–115.

Mann, R., Malisoux, L., Nuhrenborger, C., Urhausen, A., Meijer, K., and Theisen, D. (2015a). Association of previous injury and speed with running style and stride-to-stride fluctuations. *Scandinavian Journal of Medicine & Science in Sports*, 25(6):e638–645.

Mann, R., Malisoux, L., Urhausen, A., Statham, A., Meijer, K., and Theisen, D. (2015b). The effect of shoe type and fatigue on strike index and spatiotemporal parameters of running. *Gait & Posture*, 42(1):91–95.

Napier, C. and Willy, R. (2018). Logical fallacies in the running shoe debate: let the evidence guide prescription. *British Journal of Sports Medicine*, 52(24):1552–1553.

Neal, B., Griffiths, I., Dowling, G., Murley, G., Munteanu, S., Franettovich Smith, M., Collins, N., and Barton, C. (2014). Foot posture as a risk factor for lower limb overuse injury: a systematic review and meta-analysis. *Journal of Foot and Ankle Research*, 7(1):55.

Nielsen, R., Bertelsen, M., Ramskov, D., Damsted, C., Brund, R., Parner, E., Sorensen, H., Rasmussen, S., and Kjaergaard, S. (2019a). The garmin-runsafe running health study on the aetiology of running-related injuries: rationale and design of an 18-month prospective cohort study including runners worldwide. *BMJ Open*, 9(9):e032627.

Nielsen, R., Bertelsen, M., Ramskov, D., Moller, M. Hulme, A., Theisen, D., Finch, C., Fortington, L., Mansournia, M., and Parner, E. (2019b). Time-to-event analysis for sports injury research part 1: time-varying exposures. *British Journal of Sports Medicine*, 53(1):61–68.

Nielsen, R., Bertelsen, M., Ramskov, D., Moller, M. Hulme, A., Theisen, D., Finch, C., Fortington, L., Mansournia, M., and Parner, E. (2019c). Time-to-event analysis for sports injury research part 2: time-varying outcomes. *British Journal of Sports Medicine*, 53(1):70–78.

Nielsen, R., Buist, I., Parner, E., Nohr, E., Sorensen, H., Lind, M., and Rasmussen, S. (2014). Foot pronation is not associated with increased injury risk in novice runners wearing a neutral shoe: a 1-year prospective cohort study. *British Journal of Sports Medicine*, 48(6):440–447.

Nielsen, R., Buist, I., Sorensen, H., Lind, M., and Rasmussen, S. (2012). Training errors and running related injuries: a systematic review. *International Journal of Sports Physical Therapy*, 7(1):58–75.

Nielsen, R., Malisoux, L., Moller, M., Theisen, D., and Parner, E. (2016). Shedding light on the etiology of sports injuries: a look behind the scenes of time-to-event analyses. *Journal of Orthopaedic & Sports Physical Therapy*, 46(4):300–311.

Nigg, B. (2001). The role of impact forces and foot pronation: a new paradigm. *Clinical Journal of Sport Medicine*, 11(1):2–9.

Nigg, B., Bahlsen, H., Luethi, S., and Stokes, S. (1987). The influence of running velocity and midsole hardness on external impact forces in heel-toe running. *Journal of Biomechanics*, 20(10):951–959.

Nigg, B., Baltich, J., Hoerzer, S., and Enders, H. (2015). Running shoes and running injuries: mythbusting and a proposal for two new paradigms: 'preferred movement path' and 'comfort filter'. *British Journal of Sports Medicine*, 49(20):1290–1294.

Oja, P., Titze, S., Kokko, S., Kujala, U., Heinonen, A., Kelly, P., Koski, P., , and Foster, C. (2015). Health benefits of different sport disciplines for adults: systematic review of observational and intervention studies with meta-analysis. *British Journal of Sports Medicine*, 49(7):434–440.

Perkins, K., Hanney, W., and Rothschild, C. (2014). The risks and benefits of running barefoot or in minimalist shoes: a systematic review. *Sports Health*, 6(6):475–480.

Richards, C., Magin, P., and Callister, R. (2009). Is your prescription of distance running shoes evidence-based? *British Journal of Sports Medicine*, 43(3):159–162.

Ryan, M., Elashi, M., Newsham-West, R., and Taunton, J. (2014). Examining injury risk and pain perception in runners using minimalist footwear. *British Journal of Sports Medicine*, 48(16):1257–1262.

Ryan, M., Valiant, G., McDonald, K., and Taunton, J. (2011). The effect of three different levels of footwear stability on pain outcomes in women runners: a randomised control trial. *British Journal of Sports Medicine*, 45(9):715–721.

Saragiotto, B., Yamato, T., Hespanhol Junior, L., Rainbow, M., Davis, I., and Lopes, A. (2014a). What are the main risk factors for running-related injuries? *Sports Medicine*, 44(8):1153–1163.

Saragiotto, B., Yamato, T., and Lopes, A. (2014b). What do recreational runners think about risk factors for running injuries? A descriptive study of their beliefs and opinions. *Journal of Orthopaedic & Sports Physical Therapy*, 44(10):733–738.

Squadrone, R. and Gallozzi, C. (2009). Biomechanical and physiological comparison of barefoot and two shod conditions in experienced barefoot runners. *The Journal of Sports Medicine and Physical Fitness*, 49(1):6–13.

Squadrone, R., Rodano, R., Hamill, J., and Preatoni, E. (2015). Acute effect of different minimalist shoes on foot strike pattern and kinematics in rearfoot strikers during running. *Journal of Sports Sciences*, 33(11):1196–1204.

BIBLIOGRAPHY

Stacoff, A., Reinschmidt, C., Nigg, B., van den Bogert, A., Lundberg, A., Denoth, J., and Stussi, E. (2000). Effects of foot orthoses on skeletal motion during running. *Clinical Biomechanics*, 14(1):54–64.

Tam, N., Astephen Wilson, J., Noakes, T., and Tucker, R. (2014). Barefoot running: an evaluation of current hypothesis, future research and clinical applications. *British Journal of Sports Medicine*, 48(5):349–355.

Theisen, D., Malisoux, L., Genin, J., Delattre, N., Seil, R., and Urhausen, A. (2014). Influence of midsole hardness of standard cushioned shoes on running-related injury risk. *British Journal of Sports Medicine*, 48(5):371–376.

Theisen, D., Malisoux, L., Gette, P., Nührenbörger, C., and Urhausen, A. (2016). Footwear and running-related injuries – running on faith? *Sports Orthopaedics and Traumatology Sport-Orthopädie-Sport-Traumatologie*, 32(2):169–176.

van der Worp, H., Vrielink, J., and Bredeweg, S. (2016). Do runners who suffer injuries have higher vertical ground reaction forces than those who remain injury-free? A systematic review and meta-analysis. *British Journal of Sports Medicine*, 50(8):450–457.

van Gent, R., Siem, D., van Middelkoop, M., van Os, A., Bierma-Zeinstra, S., and Koes, B. (2007). Incidence and determinants of lower extremity running injuries in long distance runners: a systematic review. *British Journal of Sports Medicine*, 41(8):469–480.

Videbaek, S., Bueno, A., Nielsen, R., and Rasmussen, S. (2015). Incidence of running-related injuries per 1000 h of running in different types of runners: A systematic review and meta-analysis. *Sports Medicine*, 45(7):1017–1026.

Withnall, R., Eastaugh, J., and Freemantle, N. (2006). Do shock absorbing insoles in recruits undertaking high levels of physical activity reduce lower limb injury? A randomized controlled trial. *Journal of the Royal Society of Medicine*, 99(1):32–37.

Yamato, T., Saragiotto, B., and Lopes, A. (2015). A consensus definition of running-related injury in recreational runners: a modified Delphi approach. *Journal of Orthopaedic & Sports Physical Therapy*, 45(5):375–380.

Chapter 7

Markov Chain Modelling and Simulation in Net Games

MARTIN LAMES
TECHNISCHE UNIVERSITÄT MÜNCHEN

Abstract

Net games like tennis or table tennis may be structured as a sequence of rallies from service to error or winner. A rally can be perceived as an alternating sequence of strokes, each taking place in a certain class of strokes, e.g., base line stroke player A. This structure makes net games suited for state-transition modelling using finite Markov chains. Transition matrices for matches are obtained by game observation and provide a description of the match as a "super-rally". Performance indicators may be calculated from this matrix and used for empirical validation of assumptions made when treating it as a Markov chain. Moreover, one may then simulate game behaviour and study the impact of these changes on overall success, i.e., the point winning probabilites. This can be done by direct changes in the transition probabilities or by analysing partial or directional derivatives obtained by numerical differentiation. The full potential of Markov chains for modelling net games and other sports is not exploited yet, especially employing higher-order, continuous, or drifting Markov chains seems to be well-suited for generating further insights in many sports.

7.1 Introduction

Markov processes are universal tools in mathematics and stochastics. They are used in many applications in various fields like queueing theory (Gross et al., 2008) and genome analyses (Vergne, 2008). This chapter introduces the application of special Markov processes, discrete finite absorbing Markov chains, to performance analysis (PA) of net games like tennis or table tennis. We will see that they are not only good descriptions for net games, but they might as well be used for answering practical questions of PA using simulations.

The basic idea of finite Markov chains is to describe a process by a finite set of discrete states and the transition probabilities between them. If the Markov property - as discussed below - may be assumed, many interesting variables can be obtained by rather simple calculations using the transition matrix between states. What makes this model appealing to sports science is that the variables obtained are of high relevance in PA like winning probabilities or rally lengths.

The basic idea of this modelling approach is to perceive a rally in a net game as a sequence of states. In tennis, for example, we have the starting states *Service player A* and *Service player B*. The rally typically continues with the second stroke, *Return B* or *Return A*, with observable transition probabilities. In case of a service error the second states of a rally are *Second service A* or *B*. Another possible transition of a *First Service A/B* is an ace, i.e., a direct point scored with the serve. In the model language this is a transition from state *Service A/B* to the state *Point A/B*. This introduces an interesting type of states widely used in Markov chains, the absorbing states, i.e. states from which the process cannot escape. Obviously, the two possible outcomes of a rally, *Point for A/B*, may quite naturally be described as absorbing states.

Moreover, having a mathematical model allows, in principle, for simulations, i.e., one may manipulate transition probabilities and study the effect of these changes on absorption/winning probabilities. Thus, one may give answers to typical questions of PA, e.g., how does a change in a certain behaviour affect my overall success, what is the most efficient measure to improve my overall performance, what are the most important targets for training, and to which targets should I devote my resources in training?

This chapter starts with an explanation of the mathematical properties of finite absorbing Markov chains and on their applicability to net games. Methods for simulation are introduced and results are reported. Finally, an outlook is given on how to improve practical impact of the modelling and on the use of more sophisticated variants of Markov chains.

7.2 Markov Chain Models

Model building and simulation are standard methods in science. Their principles and methods have become the topic of many books and papers addressing more general questions like the nature of the model relation that connects original and model (Stachowiak, 1973), types and levels of modelling (Spriet and Vansteenkiste, 1982) and model validation (Perl et al., 2002). One prerequisite for successful modelling is to understand the nature and properties of the model and how far these reflect properties of the original, here, net games. This paragraph analyses properties of finite absorbing Markov chains with respect to their applicability to net games.

7.2.1 Markov Chains

The definitions and formal relationships in this paragraph are taken from the textbook "Finite Markov Chains" by Kemeny and Snell (1976).

A Markov process is a stochastic process that satisfies the Markov property (discussed below). A finite Markov process is a Markov process with a finite number of states. There are slightly different opinions in the literature on what constitutes a Markov chain. Kemeny and Snell (1976) state that a Markov chain is a Markov process with transition probabilities independent of n, which is the position of the states in their sequence. This could be called time invariance where the process steps are given by discrete time intervals.

The most important feature of Markov chains, though, is the Markov property.

7.2.2 The Markov Property

The most common metaphorical paraphrase of the Markov property is "memorylessness". That means that the probability of the process being in state s_i in the n-th step depends only on the state it was in in the step before: s_j in step $n-1$ (where $i, j = 1, \ldots, k$ and k = number of states). Using the outcome function f that gives the state s_i of the process at each step $n(n = 0, 1, 2, \ldots)$ the Markov property is defined as:

$$\begin{aligned} &\mathrm{P}(f_n = s_{i,n} | f_{n-1} = s_{j,n-1}, f_{n-2} = s_{j,n-2}, \ldots, f_0 = s_{j,0}) \\ &= \mathrm{P}(f_n = s_{i,n} | f_{n-1} = s_{j,n-1}). \end{aligned}$$

In other words: The conditional probability that the process is in state s_i in the n-th step given all the states before n $(0, \ldots, n-1)$ is the conditional probability of being in state s_i in the n-th step given only the state before n. Another paraphrase: the future depends only on the present and not on the past.

As a first consequence of the Markov property, a Markov chain is determined entirely by an initial probability vector of states (starting vector π_0) and its transition matrix P with transition probabilities $p_{i,j}$ denoting the probability of a transition from state s_i to state s_j:

$$\pi_n = \pi_0 \bullet P^n.$$

That means we can obtain the state vector π_n of state n by multiplying the starting vector π_0 n times with the transition matrix.

In the case of absorbing Markov chains, we can use this property to calculate the absorbing probabilities given a certain starting vector iteratively: the starting vector is repeatedly multiplied by the transition matrix until the probability of the process not being caught in an absorbing state falls below a certain threshold . When = 0.001 is assumed in tennis, for example, we may assume that less than 0.1% of the rallies are still running, which means that the error of the absorbing probabilities is negligible.

The validation of the Markov property is subject to debate in the literature (Bickenbach and Bode, 2001). Autocorrelation studies are recommended but rare in practice. Thus, the strategies of content validation (i.e., the discussion whether there are objections against the assumption of the Markov property) and predictive validation (i.e., the empirical testing of predictions assuming the Markov property) are most common and will be applied to net games below.

7.2.3 Finite Markov Chains

Finite Markov chains constitute a class of stochastic processes with certain specific properties, making them very appropriate, simple, and useful models for net games and many other real-life processes (examples in Kemeny and Snell (1976)). It has become common to distinguish between certain types of states: starting states, transient states, and absorbing states.

Starting states are possible states for $n = 0$, i.e., states which the process under consideration may start with. As mentioned above, in the case of net games, these are the service of player A and B. Typical starting vectors in net game studies are 0 for all states except the services. They may either contain an initial probability of 0.50 for service player A and B (appropriate for table tennis because of alternating service) or the empirical relative frequencies of services from A and B (more appropriate for tennis because of *a priori* unknown number of services). If only rallies with service from either A or B are to be studied, they are given a starting probability of 1.

Transient states are states to which the process may never return

7.2. MARKOV CHAIN MODELS

once this state has been left. The same holds for transient sets of states; once this set of states is left, the process may never return to it. Examples in tennis are *Service* or *Return* for transient states: there is no way to return to the state *Service* in a tennis rally. In table tennis long rallies end up in the transient set of states > 4 *stroke A* and > 4 *stroke B* unless the rally is absorbed in *Point for A* or *Point for B*.

Absorbing states are most interesting not only from a sports point of view. The formal definition of an absorbing state, $p_{i,i} = 1$, expresses quite clearly: once entered in an absorbing state, the process is trapped; it will remain forever in this state. In net games these absorbing states are the ultimate aim for each player in a rally, making it end up in *Point A* or *Point B*.

The theory of finite Markov chains allows calculating some variables that are of general as well as sport specific interest.

If we have r absorbing and s transient states, we can give a general transition matrix of a finite absorbing Markov chain as follows:

$$P = \begin{bmatrix} I_r & 0 \\ R & Q \end{bmatrix},$$

with

- I_r being the $r * r$ unit matrix of the transitions between r absorbing states ($r = 2$ in net sports: *Point for A* and *Point for B*),
- *0* being the $r * s$ matrix with all $p_{i,j} = 0$ because of the r absorbing states,
- R being the $s * r$ matrix with the transition probabilities from the transient states to the absorbing states, and
- Q being the $s * s$ matrix with the transition probabilities between the transient states.

A "fundamental matrix" N (Kemeny and Snell, 1976) may be obtained as follows:

$$N = (I_s - Q)^{-1},$$

with I_s being the $s * s$ unit matrix. With the help of N, one may calculate the following variables:

1) The entries $n_{i,j}$ in this matrix represent the *expected frequencies* of the process touching state j when being in state i. Moreover, matrix N_{var} gives the variance of each $n_{i,j}$ of N:

$$N_{var} = N * (2 * N_{dg} - I_s) - N_{sq},$$

with N_{dg} being the diagonal of matrix N and N_{sq} being the matrix of all $n_{i,j}^2$. This allows calculating confidence limits for each $n_{i,j}$ that can be used for statistical tests testing whether observed frequencies are in agreement with predictions assuming that a Markov chain is appropriate.

2) Also, one may calculate the *expected number of steps* from a state to any absorbing state. This number is just the vector *tau* of the sums of the rows in N:

$$tau_i = \sum_{j=1}^{s} n_{i,j}.$$

Again, the variance may be calculated:

$$tau_{var} = (2 * N - I_s) * tau - tau_{sq},$$

with tau_{sq} bearing the same explanation as N_{sq}.

3) The *absorption probabilities* in any of the absorbing states, starting at a certain state are given by matrix multiplication:

$$B = N * R,$$

with N being the "fundamental matrix" and R from the partitioning of P given above. B is an $s * r$-matrix with entries $b_{i,j}$ giving the probability of being absorbed in absorbing state j once transient state i is reached. This allows calculating the general point winning probabilities when starting in *Service A* or *Service B*. More detailed absorbing probabilities are also of interest: if you manage to arrive at base-line game after service and return, how big is your chance of winning the rally now?

The question of validation was already addressed above: how is it possible to make sure that we may see a tennis rally as a finite Markov chain? Besides considerations that focus on the nature of the processes (net games), the theoretical results presented above make it possible to compare predicted values of the process (frequencies in certain states, number of steps until absorbed, transition probabilities to absorbing states) to observed quantities in the sense of empirical validation.

The "chain" property, i.e., the invariance of the process over time, may be validated in a straightforward manner by dividing the process in different parts and demonstrating only random deviations between the transition matrices of these different parts, e.g., sets in tennis.

7.3 Net Games as Finite Markov Chains

In this section the relation between the original, net games, and the model, finite Markov chains, is established and discussed. After some general structural remarks on net games from the perspective of sports science, it is shown how the answers for typical questions of PA may be obtained by Markov chain modelling. After this, validation is addressed focusing on content validation and empirical results.

7.3.1 State-Transition Modelling of Net Games

Net games are sports with two opponents (single or double) played on a symmetrical court separated by a net (exception: squash). The ball (or object of game like shuttlecock in badminton) is played with a racket (exceptions: US-handball, peteca/indiaca: played with hand). Presently, tennis, table tennis, and badminton are net games that are part of the Olympic sports program.

Net games consist of sets (tennis: sets and games) that are won when a certain number of points is scored. These points are won by winning a rally, i.e., a series of alternating strokes starting with a service and ending with a point for player A or B. A point is awarded either when a player hits a stroke that cannot be answered by a stroke from the opponent (winner) or when the opponent hits an error (ball/game object in net or out of field).

This particular structure as series of alternating strokes from different categories (service, return, ...) gives rise to modelling net games with state-transition models (see Figures 7.1 and 7.2 for state-transition models for tennis and table tennis). Each stroke category makes up two discrete states (one for each player) with the absorbing states *Point for A or B* added. The transitions between these states are given by the relative frequencies of successor states for each state and may by depicted by a transition matrix (see Tables 7.1 and 7.2 for transition matrices for tennis and table tennis matches). These transition probabilities may be obtained by observational methods from real matches (Lames, 1994).

State-transition modelling is applicable to net games for several reasons. The alternating strokes take place in classes of situations that give rise to equivalence classes of states of each net game in a very natural way. We have clear transitions: the crossing of the net. Dynamics and interaction, the constitutive elements of game sports (net sports and team sports), are contained here as both players want to reach their favourable absorbing state with each stroke being devoted to this task.

Figure 7.1 shows the state-transition model for tennis rallies and Figure 7.2 for table tennis rallies. In tennis we have the starting state

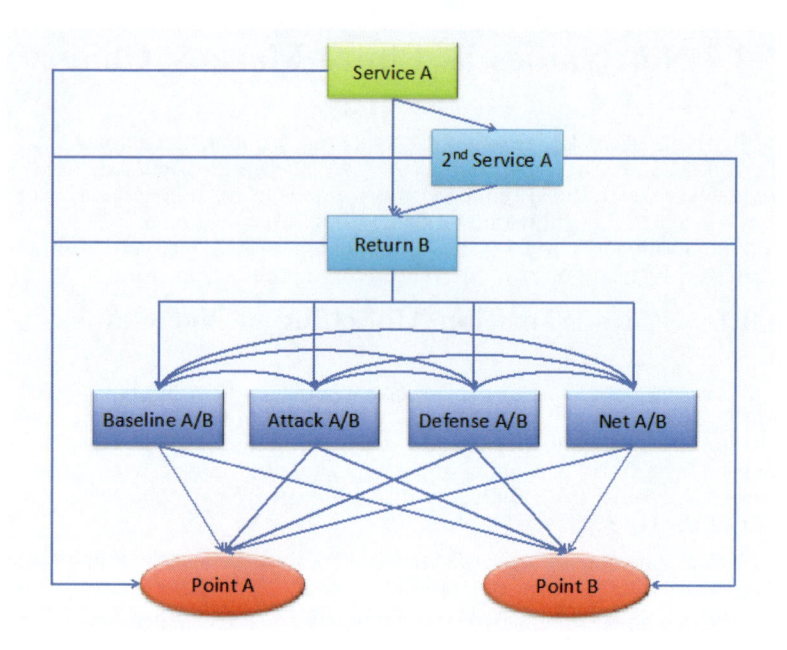

Figure 7.1: State-transition models for tennis.

Service A (states for player B serving are symmetrical). Then, the two transient states follow: *2nd Service A* and *Return B*. After the opening of the rally, a transient set of states is entered. This set contains four states of open play, baseline strokes, attack strokes (player at net, opponent at baseline), defence strokes (player at baseline, opponent at net), and net strokes (both players at net). The arrows between the states of this transient set make clear that, at least in principle, there may be a transition from each state to every other state, e.g., if a lob is played in the net state the match may continue with a defence stroke (lobbed player at baseline, lobbing player stays at net) or even a baseline stroke (both players at baseline). The absorbing states *Point A* and *Point B* may be reached from each state, except *Point B* from *Service A*.

The state-transition model for *table tennis* is related to the number of the strokes. Implicitly, we have tactical meanings with the first stroke being the service, the second stroke being the return. The depicted model is based on the modelling of the Chinese Table Tennis Federation (Fuchs et al., 2018) where 1st and 3rd as well as 2nd and 4th stroke are important stroke classes made complete by longer rallies with more than four strokes.

7.3. NET GAMES AS FINITE MARKOV CHAINS

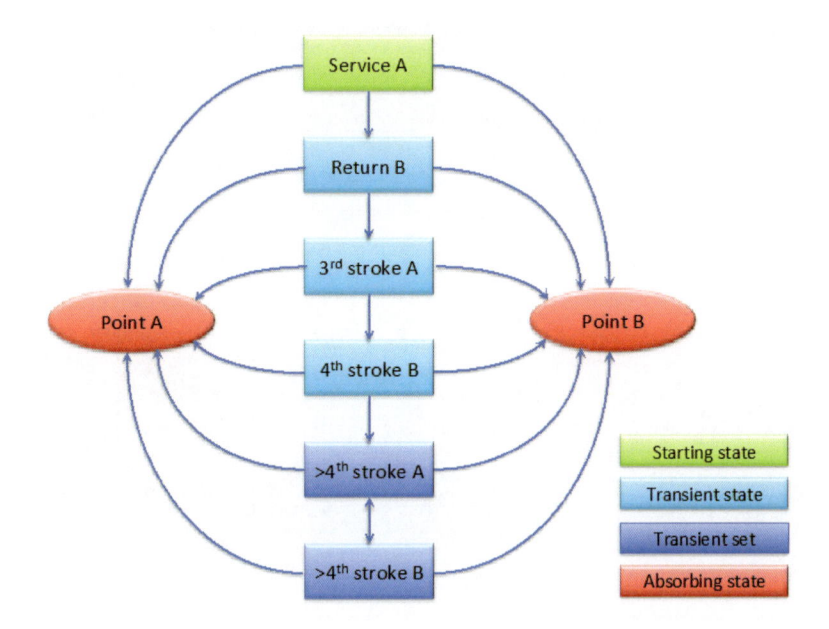

Figure 7.2: State-transition models for table tennis.

From a mathematical perspective it is interesting to note that the chain property of Markov processes is acknowledged here in a way that earlier strokes in a rally have specific transition probabilities, whereas after four strokes these probabilities may be assumed to be invariant.

Figures 7.1 and 7.2 depict examples for transition matrices from real matches in tennis and table tennis, respectively. These give precise descriptions of the match at the chosen level of abstraction. The transition matrix describes a "super rally", including each single rally in its transition probabilities. For sports experts it is quite obvious from the transition matrix what was going on and where the winner took advantage most.

In the tennis matrix in Table 7.1 we see that Jennifer Capriati showed a more aggressive service with more first service errors and double faults. Obviously, this provoked the high return error rate of Kim Clijsters. Moreover, it is obvious that the most prevalent course of the rallies was the baseline duel with only low transition rates to the net, exclusively from baseline strokes meaning that both players never played serve-and-volley. Finally, we see that the match was very tight because there is no obvious advantage for the winner in the transition probabilities.

Table 7.1: Transition matrix for tennis: Clijsters vs. Capriati, Australian Open Half Final, 2002; 7-5/3-6/6-3.

Clijsters	A Serv2	B Ret	B Base	B Att	B Def	B Net	Point A	Point B
1st Serv	32.4	66.2					1.4	
2nd Serv		91.3					0.0	8.7
Return			85.3	0.0	0.0	0.0	0.0	14.7
Baseline			80.9	2.6	0.6	0.0	1.4	14.6
Attack			0.0	0.0	62.5	0.0	37.5	0.0
Defence			20.0	40.0	0.0	0.0	0.0	40.0
Net			0.0	0.0	0.0	0.0	100.0	0.0

Capriati	B Serv2	A Ret	A Base	A Att	A Def	A Net	Point A	Point B
1st Serv	42.9	53.6						3.5
2nd Serv		83.3					16.7	0.0
Return			89.4	0.0	0.0	0.0	7.6	3.0
Baseline			83.1	1.7	1.4	0.0	11.7	2.0
Attack			0.0	0.0	35.7	14.3	14.3	35.7
Defence			25.0	25.0	0.0	0.0	50.0	0.0
Net			0.0	0.0	0.0	0.0	0.0	0.0

In contrast to this tennis match, the dominance of the winner becomes quite clear in the table tennis example (Table 7.2). Whereas in service and return Ma Lin shows a weaker result committing more errors, his strong strokes are # 3 and # 4 with longer rallies being more balanced again. The dominance here allowed for a clear victory as most of the rallies finished in this phase.

7.3.2 Performance Analysis Using Markov Chains

Besides the informative but purely descriptive use of the transition matrices, one may obtain interesting performance indicators from variables calculated assuming the properties of Markov chains.

The expected rally length given an arbitrary starting state depicts an important aspect of tactics, i.e., the capability to maintain the rally after a certain point or to finish it when this is more favourable. Similar things may be found out analysing the absorption probabilities from each state. These probabilities give a more complete picture about the dominance in a certain state than the transition probabilities of this state alone because it includes delayed effects, too.

Taking rally length and absorption probability for each state together, this informs about the tactical battle in a match. As we saw in the table tennis example, there are advantageous states for a player. The performance indicators above show whether a player was able to conduct the rally into these advantageous states or prevent going in the disadvantageous ones, for example, Ma Lin being able to finish

7.3. NET GAMES AS FINITE MARKOV CHAINS

Table 7.2: Transition matrix for table tennis: Ma Lin vs. Wang Hao, Olympic Final, 2008; 11-9/11-9/6-11/11-7/11-9.

Ma Lin	Ret B	#3 B	#4 B	#5 B	> 5 B	Point A	Point B
Service	97.9					0.0	2.1
Return		91.5				0.0	8.5
Stroke #3			88.9			4.4	6.7
Stroke #4				79.3		6.9	13.8
Stroke #5					73.3	3.3	23.3
Stroke > 5					80.9	0.0	19.1

Wang Hao	Ret A	#3 A	#4 A	#5 A	> 5 A	Point A	Point B
Service	100.0					0.0	0.0
Return		95.7				4.3	0.0
Stroke #3			67.4			23.3	9.3
Stroke #4				75.0		20.0	5.0
Stroke #5					65.2	26.1	8.7
Stroke > 5					67.0	26.0	6.8

the rally before reaching disadvantageously longer rally lengths.

7.3.3 Model Validation

As mentioned above, the validation of the general assumptions of finite Markov chains, i.e., the Markov property and time invariance, is prerequisite for further analyses making use of the calculations cited above. Especially using finite Markov chains for simulations as shown in the next paragraph makes use of these assumptions extensively. Nevertheless, validation of Markov chain properties is often neglected (Bickenbach and Bode, 2001).

Model validation in practice consists of testing for content validity and predictive validity. Content validity is the result of an expert discussion ("face validity") or based on logical arguments and will here be explained for tennis.

First, one must acknowledge that the structure of net games allows comprehensibly for state-transition modelling. Figure 7.1 depicts states for tennis that are widely accepted in sports practice. Also, the specification of starting states, transient states, transient sets of states, and finally absorbing states reflects rallies in net games very directly.

Nevertheless, there are some objections against stationarity. It is well known in practice and has been studied and demonstrated in table tennis (Fuchs et al., 2018) that we find phases where players show streaks of higher and lower success in a match. Moreover, one might assume different transition probabilities for base line strokes in tennis between earlier and later base line strokes in a rally due to accumulated fatigue.

One might argue against the assumption of the Markov property itself, too. For example, there is good reason to assume that transition probabilities in state *Return* differ to a great extent, depending on whether the state before was *First service* or *Second service*. We know of different service as well as return tactics for first and second services. This would be a clear violation of the Markov property, which demands that the probability of the next state does not depend on the past, e.g., the probability of state *Point for opponent* after a return may not differ between returns after a first and second service. (By the way, this specific objection could be removed by introducing a new state *Return on 2nd Service*).

As we have seen, there are good reasons from a sports practice point of view against the assumptions of finite Markov chains. But how is it possible to judge whether these objections are severe enough to prohibit finite Markov chain modelling? Here, predictive validity comes into play. There are numerous variables that may be calculated making use of these presumed properties. Typically, these variables may be obtained by game observation, too. For example, the well-known match statistic of overall point winning probability when serving is predicted by the absorbing probability starting in state *First Service A* or *First Service B*. Both variables may be easily compared and when comparisons reveal good agreement, this is a good argument that the assumptions of finite Markov chains are not violated in a too heavy way to allow for good predictions.

Moreover, for some variables like expected frequency of a state in a rally and remaining rally length starting from any state, the Markov chain theory provides estimates for the variance of these variables. This allows calculating confidence intervals and, based on this, significance testing (assuming a normal distribution). The results in each case where predictions have been compared to the observed values so far state unanimously that deviations between observed values and values calculated under Markov chain assumptions differed only marginally. Typically, the overall winning probability is missed by less than 0.01.

For example, the largest examination of predictive validity so far (Lames, 1991) based on confidence intervals on match base resulted in very acceptable predictions. For 306 players in 153 tennis matches the winning probabilities, rally lengths and state frequencies of eight non-trivial states (like frequency 1 for *First service A*) were calculated with their 95% confidence interval. In only six out of the 306 cases the observed values were found outside.

More recently, Wenninger and Lames (2016) tested the prediction for point winning probabilities for 518 players in 259 table tennis matches. They found a mean difference to the observed values of 0.003% with a standard deviation of 0.118%. In table tennis the assumption of Markov chain properties for a rally leads to even higher accurate predictions of point winning probabilities than in tennis. This is due to the more complicated state structure in tennis, especially with the transient set of four states in longer rallies.

To sum up, one must admit, despite substantial objections against the Markov property as well as stationarity, empirical validations of predictions show a high agreement so far without a reported exception. This may be taken as a justification to make use of finite Markov chains as models for net games.

7.4 Game Simulations Based on Finite Markov Chains

In general, simulation is defined as the creation of a mathematical model to imitate the behaviour of a system for a certain purpose (Onpulson.de, 2019). In sports, especially in net games, the questions one wants to answer with simulation differ between theoretical and practical performance analysis (Lames and McGarry, 2007). In practical PA one is interested in questions such as "What would have happened if my player would have committed 5% less service or return errors or would have played more aggressively in base line rallies and attacked more often at the net?" Theoretical PA is interested in general laws and in sensitivity analyses such as "Which transition has the most impact on point winning probability? Which transitions are indifferent?".

If one is able to give answers to these questions many interesting further questions may be answered such as "What are the most important targets for training? Which behaviours have the best cost-benefit relation for training?". The perspective to get answers for these central questions of performance analysis already incited early simulations with finite Markov chain models of net games.

7.4.1 Simulations Using Changes in Transition Probabilities

The procedure of simulations using finite Markov chains is straightforward. To study the impact of a behaviour change, e.g., 5% less service errors, one may follow a three-step approach:

1. Calculate the point winning probability of player A using the observed transition matrix.

2. Simulate the behaviour under scrutiny by manipulating the transition matrix accordingly. In our example this would mean to decrease the transition probability from state *First Service A* to *Second Service A* by 5% (Note: as the sum of all transitions of

First Service A must be 100% , this decrease must be compensated).

3. Now, calculate the point winning probability of player A again. The difference to the initial point winning probability may be interpreted as the impact of this behavioural change on overall winning probability.

There may be different strategies for compensating the changes reflecting the tactical behaviour under scrutiny. If there is a specific suggestion, for example, a practitioner may state very comprehensively that a decreased error rate is compensated by an increased transition to *Return B* and not by an increase in *Point A* (aces), this may be implemented as well as more general models, for example, a compensation by increasing all the other transition probabilities in sum by 5% in proportion to their size.

If we want to compare the impact of different behavioural changes on overall point winning probability, which is a typical question in PA, we need another modelling step. We need to change the initial transition probabilities to an extent that should be of comparable difficulty, i.e., the difficulty of a behavioural change in a net game must be modelled. The nearby way of a constant percentage is doubtful. There is good reason to assume that it is less difficult to reduce the return error rate from 30 to 25% than to reduce the rate of double faults from 6 to 1% .

Lames (1991) suggested a model for comparable changes of transition probabilities as a sum of a constant change that is possible even in extreme areas (close to or at transitions of 0 and 100%) and a relative change that is proportional to the distance to either 0% or 100% because changing behaviour close to the boundaries may be assumed to be more difficult than far away from them. This results in a formula for calculating comparable changes in transition probabilities ($\Delta p_{i,j}$):

$$\Delta p_{i,j} = K + 4B * p_{i,j} * (1 - p_{i,j}),$$

with K being a basic constant change that is possible independent of the level of $p_{i,j}$ and B denoting the maximum relative change at $p_{i,j}$ =0.5. The constant 4 makes sure that B is realised as the maximum of $p_{i,j} * (1 - p_{i,j})$ is 0.25. $\Delta p_{i,j}$ is added to the original transition probability if $p_{i,j} < 0.5$ and subtracted if $p_{i,j} \geq 0.5$.

With a representative sample of net games, one is now able to simulate several relevant tactical behaviours. (Note: these must be expressed with changes in transition probabilities.) This gives rise to an ordering of behaviours according to their impact on winning probability, which is a central question of theoretical PA. Also, one may study the importance of behaviours in dependence of the gender of

7.4. GAME SIMULATIONS

players or the court surface (in tennis) or other independent variables like performance level, handedness, or player's nation. The dependent variable is the per cent change of the winning probability when changing transitions describing a certain behaviour to the degree mentioned above and compensating this change. It may be called the relevance of the tactical behaviour under scrutiny.

In tennis (Lames, 1991) it was found that the baseline error is in general the most relevant transition, and only on faster pitch surfaces like lawn, where for example a service does not lose as much speed when hitting the surface as on clay, the return error is more important. Figure 7.3 shows the mean values of relevance of the baseline error for clay, artificial turf, and lawn surfaces for men and women of top level (world ranking < 80). The factor gender is significant because men finish many rallies with service and return, making baseline transitions less important. The factor surface is expectedly influential as baseline game is more important on slower courts. And finally, there is a significant interaction because there is almost no difference between genders on clay, but there are large differences on faster surfaces.

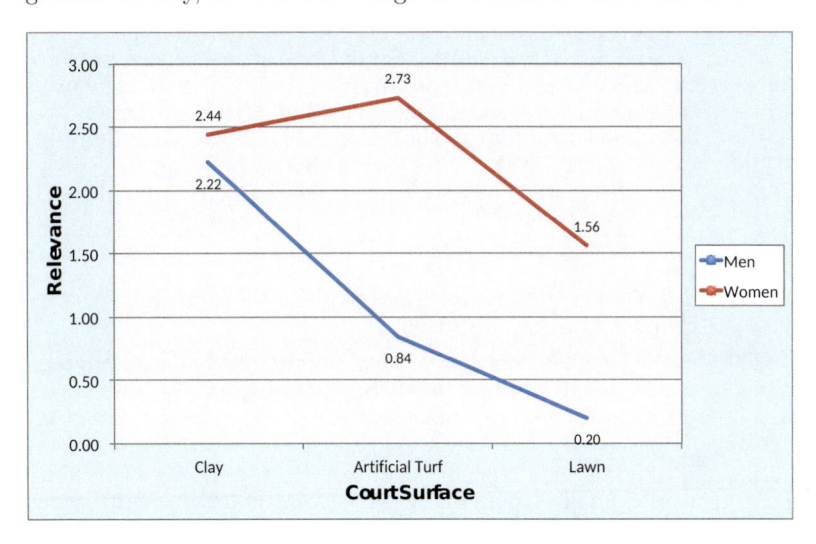

Figure 7.3: The relevance of baseline errors on different surfaces for men and women.

7.4.2 Simulations Using Numerical Differentiation

A possible improvement of the above-mentioned method for simulation was suggested by Wenninger and Lames (2016). It omits the necessity of modelling comparable difficulties of tactical behaviours and the question of how to compensate for changed transition probabilities. Despite these advantages the suggested procedure is much more demanding numerically and still not fully validated so that for the time being, the two methods co-exist.

The basic idea of this alternative concept of simulation is to see the calculation of the point winning probabilities or absorption probabilities in Point A when starting in Service A as a function $f : R^n \rightarrow R$ with n being the number of all transition probabilities. Here, the point winning probability is conceived as a landscape or potential over an n-dimensional parameter space. Applying the transition matrices presented in Tables 7.1 and 7.2, we have $n=72$ in tennis and $n=36$ in table tennis.

Now, the relevance of some tactical behaviour expressed in a transition probability, i.e., a dimension of the parameter space, may be seen as the gradient of the point winning probability surface in the direction of the transition probability. If there is a steep increase, the winning probability is very sensitive to this behaviour; if the gradient is around zero, this behaviour is only slightly influential.

As there is no analytical solution for obtaining the derivatives from the necessary calculations, numerical differentiation is used. The procedure obtains a numerical solution for the differentiation equation, respectively for the central differentiation equation (Burden and Faires, 2010):

$$f'(x) = \lim_{h \to 0} \frac{f(x+h) - f(x)}{h} \text{ resp. } f'(x) = \lim_{h \to 0} \frac{f(x+h) - f(x-h)}{2h},$$

with x being the vector of transition probabilities and h being the vector of changes of these transition probabilities.

The limit approximation in numerical differentiation is carried out by using a very small h which must be chosen carefully because of rounding errors and computer inaccurateness. A suggestion found in literature (Press et al., 2007) is $h_i = \sqrt[4]{x_i * eps}$ with h_i $(i = 1, \ldots, n)$ given separately for each transition probability x_i and eps ("machine eps") being the smallest possible floating point number on the computer.

Working with *partial differentiation* allows for assessing the relevance of behaviours that may be described with one single transition such as error and winner rates. Table 7.3 gives results for a large, representative sample in table tennis (Wenninger and Lames, 2016). 259 singles matches from the biggest events between 2011 and 2014 were analysed containing 105 women's and 154 men's matches. One may see, for example, that error rates (trivially) always exhibit a negative impact and winner rates exhibit a positive impact on winning probability. For longer rallies ($>$ 3 strokes) results for winners and

7.4. GAME SIMULATIONS

errors are very similar (except for sign). Overall, the largest impact is found in the longest rallies, with even bigger impact for women.

Table 7.3: Partial derivatives for winners and errors in different strokes obtained by numerical differentiation and indicating the impact of the respective behaviour on winning probability.

Partial derivatives	Women		Men	
	Winner	Error	Winner	Error
Service	0.234	-0.269	0.242	-0.262
Return	0.268	-0.251	0.263	-0.260
Stroke# 3	0.228	-0.251	0.235	-0.256
Stroke# 4	0.206	-0.202	0.195	-0.192
Stroke# 5	0.153	-0.153	0.134	-0.133
Stroke# 5+	0.562	-0.557	0.347	-0.342

Besides partial differentiation one could make use of directional differentiation, modelling more complicated behaviours than just a higher single transition rate like the ones for winners and errors in Table 7.3.

For example, a higher risk-taking in a stroke class may be modelled by directional differentiation. Higher risk taking consists of an increased error probability as well as an increased winner probability in a certain stroke class. Now, one can calculate the directional derivative of error rate and winner rate combined on the high-dimensional surface of the point winning probability function. This results, like in the case of partial derivatives, in a gradient standing for the sensitivity of the winning probability against a behaviour change in this direction.

Empirical results are shown in Figure 7.4. The green dots show empirical winner and error rates for men in stroke # 4 in 154 matches. The directional derivatives for all points are calculated, and for a narrow grid Voronoi interpolations are calculated for each grid node of the convex hull. With colour coding, nodes with similar gradients are marked. These form areas with positive impact of higher risk taking (red), negative impact (blue), and neutral areas (green).

Interestingly, the results are in agreement with practical assumptions. Especially the red area at the bottom right says that players with (very) high winner probability for this class of strokes and at the same time a minimal error rate (below 2%) are well advised to increase risk, as the impact on overall winning probability is large and positive.

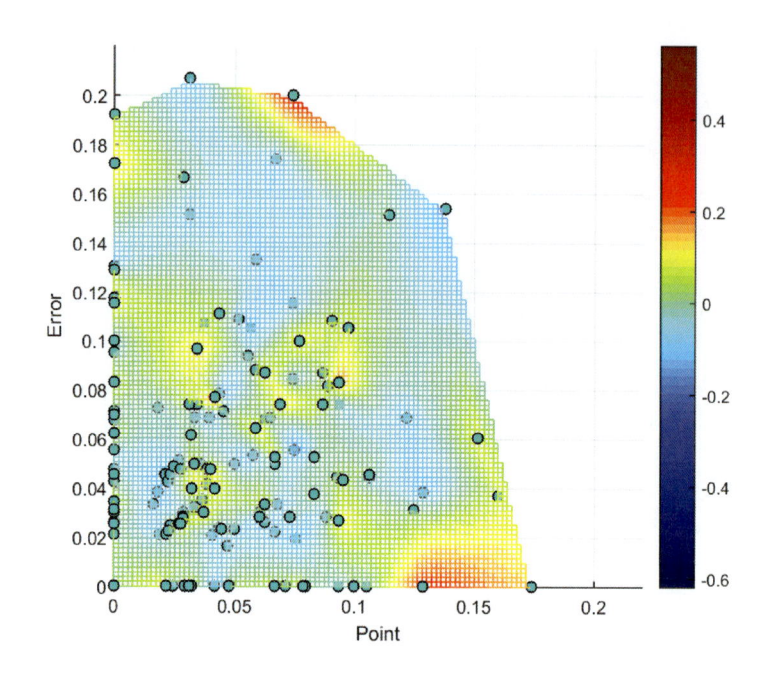

Figure 7.4: Visualisation of directional derivatives modelling the increase of risk of stroke # 4 for men.

Taken together, one may say that simulation based on Markov chains is an appropriate method in PA to give answers to central relevant questions of practical performance analysis (How effective is a 5% decrease in baseline errors, is it worth devoting more time in training to it?) and theoretical performance analysis (What is the most effective behavioural change in female top level tennis?).

The conventional simulation method introduced first must rely on estimates for changes in behaviour of comparable difficulty and to define a way for compensating the changed percentage with other transitions to preserve the 100% transitions per state. Simulations using numerical differentiation forgo these additional two modelling steps and seem to be a viable alternative. Nevertheless, this newly introduced method still needs to be confirmed from the mathematical side. For example, one should make sure that the existence of local maxima and minima frequently observed in high-dimensional functions does not question the validity of derivatives for the simulated net game behaviour.

7.5 Discussion and Outlook

7.5.1 Sports Applications

There have been a number of applications of finite Markov chains in different net games until now. The idea was introduced in tennis (Lames, 1991) quite early and transferred to table tennis by Zhang (Pfeiffer et al., 2010; Zhang, 2006; Zhang and Hohmann, 2005). Also, for the team net game volleyball, a description as transition matrix and the simulation of game behaviours using the Markov chain property were presented rather early (Lames et al., 1997). Other team sports were addressed, too. The dissertation of Pfeiffer (2005) did so for the game of handball. Finally, we find a study in football applying this method (Liu and Hohmann, 2013). One may assume that simulations using finite Markov chains could easily be extended to badminton, squash, or beach-volleyball providing valuable results on the game structure for these sports as well.

Lames et al. (1997) addressed a question of particular interest for PA in team sports using Markov chain simulation: the impact of single players. In a rather rough approach, they calculated the original winning probability from a volleyball transition matrix. Then, all transitions a single player was responsible for were eliminated from this matrix and the winning probability was determined again. The results showed, in principle, that the players who were considered to be the dominating ones got the highest ratings. Nevertheless, future efforts would need some refining as a strong influence of the number of strokes of a player was observed. Also, in volleyball we have the specific problem that the setter is part of almost each regular attack. Eliminating these contacts (sets) leaves only unusual or erroneous rallies, thus resulting in a strong bias in favour of the setter.

Table 7.4 shows an application in sports, but here a finite Markov chain does not describe a net game or an invasion game but the transitions between different performance levels for men and women in the German golf federation's talent promotion system. We have the starting state *Entry* from which most athletes enter the *C-level* but a few also at *AB-level*. The transient states are the different levels with *Re-Entry* meaning that a player dropped out of the system but managed to get in again, e.g., due to a longer injury. Without a comeback this would be the absorbing state *Dropout*. A positive way to leave the system–at least from the point of view of the athletes–in golf is to become a professional player or *Pro*.

So far only descriptions using the transition matrix have been provided, but there are interesting indicators that could be obtained in the way presented above: statistical norms for the seasons spent at a certain level, probability of becoming a pro starting at a certain level,

Table 7.4: Transition matrix for a talent promotion system in golf.

	m/f	C	AB	RE	Pro	Dropout
Entry	m	89.5	10.5			
	f	88.8	11.3			
C-Level	m	42.2	20.5	3.8	0.0	33.5
	f	45.4	22.0	1.4	0.0	31.2
AB-Level	m	1.1	60.3	1.7	8.4	28.5
	f	1.5	63.4	4.5	6.0	24.6
Re-Entry	m	0.0	64.3	28.6	7.1	
	f	0.0	44.4	55.6	0.0	

etc. These indicators could play an important role in the controlling (in the sense of business management) of the federation's talent promotion system.

7.5.2 Methodological Outlook

Compared to the valuable information obtained by simulation with finite Markov chains in sports, there are rather few applications yet. In this final section it is shown that some persisting problems could be overcome by using more appropriate models from the Markov chain theory.

In the most important sport in Europe, football, so far there is only one application (Liu and Hohmann, 2013) using transitions between different areas in the field. It is quite likely that more meaningful analyses could be done if one could model different states in ball possessions of a team (episodes). The problem with the conventional method here is that it is hard to deal with different durations of team and/or individual ball possessions. A possible solution for this problem is *continuous Markov chains* that replace the discrete time function with a continuous one, thus being much more flexible and more adapted to "continuous" sports like football, field and ice hockey, and basketball.

Some of the objections mentioned above against the Markov property could be removed when using *higher order Markov* chains. These processes modify the Markov property in the sense that the actual state does not only depend on the transitions of the last state but on the last 2, 3, ... states. This would perfectly fit to the problem of the return transitions depending on the stroke before. Second order Markov chains would, on the one hand, provide a universal solution to the problem but, on the other hand, do this at the cost of introducing many more new states. In the very small state spaces we have in modelling net or invasion games, this could also be accomplished

7.5. DISCUSSION AND OUTLOOK

by introducing additional states like *Return after 1st Service* and *Return after 2nd Service* instead of just *Return*. Nevertheless, it would be interesting to see how a higher order Markov chain would perform and maybe create even more sophisticated applications in sport.

Stationarity constitutes a hard demand for net and invasion game models as we know that there is much fluctuation of performance within matches. For example, in table tennis, Fuchs et al. (2018) analyse the momentary point probability of a player by using a moving average (double moving average of length 4) of the outcomes of points played (1=won, 0=lost). Results show considerable fluctuations within table tennis matches of elite players, typically going through almost perfect as well as disastrous phases within one match. This finding contradicts the chain property that demands a certain invariance of the performance level all over a match. In the stochastic analysis of DNA-sequences, a method was developed to get along with non-stationary processes (Vergne, 2008). *Drifting Markov chains* estimate a polynomial drift (e.g., 4th degree polynomial) and adapt the transition probabilities using this general trend. It would be interesting to apply the concept of drifting Markov chains to net or invasion games and see the differences to conventional findings.

A final objection against the presented simulations comes from sports practice. If one devotes effort in training to improve a certain capability, e.g., explosive strength, this will not only improve service speed (and ace rate as transition) but also the quality of (m)any other stroke(s). In the transition matrix, training of explosive strength might positively influence the probabilities of hitting a winner in each stroke class. The traditional method of simulation presented in this chapter assumes that a change in transition probability leaves the other transitions unchanged. The method with numerical differentiation may not be so much affected because of its limit property. Nevertheless, it would be interesting to find a *rubber-cloth-like model for the change of transition probabilities* when a change in one transition elicits changes in many (all) other transitions, too. Having such a tool at hand, we could expect to simulate the impact of an improvement in a certain aspect of performance to the overall match performance much better.

Bibliography

Bickenbach, F. and Bode, E. (2001). Markov or not Markov–this should be a question. *Kiel Working Papers, No 1086*.

Burden, R. L. and Faires, J. D. (2010). *Numerical Analysis*. Brooks/Cole, Monterey, Canada.

Fuchs, M., Liu, R., Malagoli Lanzoni, I., Munivrana, G., Straub, G., Tamaki, S., Yoshida, K., Zhang, H., and Lames, M. (2018). Table tennis match analysis: A review. *Journal of Sports Sciences*, 36(23):2653–2662.

Gross, D., Shortle, J., and Thompson, J. (2008). *Fundamentals of Queuing Theory*. Wiley, Hoboken, New Jersey.

Kemeny, J. and Snell, J. (1976). *Finite Markov Chains*. Springer, New York.

Lames, M. (1991). *Leistungsdiagnostik durch Computersimulation: Ein Beitrag zur Theorie der Sportspiele am Beispiel Tennis*. Beiträge zur Sportwissenschaft. Harri Deutsch, Frankfurt.

Lames, M. (1994). *Systematische Spielbeobachtung*. Trainerbibliothek. Philippka-Sportverlag, Münster.

Lames, M., Hohmann, A., Daum, M., Dierks, B., Froehner, B., Seidel, I., and Wichmann, E. (1997). Top oder Flop: Die Erfassung der Spielleistung in den Mannschaftssportspielen. In Hossner, E. and Roth, K., editors, *Sport-Spiel-Forschung Zwischen Trainerbank und Lehrstuhl*, pages 101–117. Czwalina, Ahrensburg.

Lames, M. and McGarry, T. (2007). On the search for reliable performance indicators in game sports. *International Journal of Performance Analysis in Sport*, 7(1):62–79.

Liu, T. and Hohmann, A. (2013). Applying the Markov chain theory to analyze the attacking actions between FC Barcelona and Manchester United in the European Champions League final. *International Journal of Sports Science and Engineering*, 7(2):79–86.

Onpulson.de (2019). https://www.onpulson.de/lexikon/simulation/.

Perl, J., Lames, M., and Glitsch, U. (2002). *Modellbildung in der Sportwissenschaft*. Hofmann, Schorndorf.

Pfeiffer, M. (2005). *Leistungsdiagnostik im Nachwuchstraining der Sportspiele : Entwicklung eines Modelltheoretischen Ansatzes im Handball*. Sport und Buch Strauß, Köln.

Pfeiffer, M., Zhang, H., and Hohmann, A. (2010). A Markov chain model of elite table tennis competition. *International Journal of Sports Science and Coaching*, 5(2):205–222.

Press, W. H., Teukolsky, S. A., Vetterling, W. T., and Flannery, B. P. (2007). *Numerical Recipes 3rd Edition: The Art of Scientific Computing*. Cambridge University Press.

Spriet, J. A. and Vansteenkiste, G. C. (1982). *Computer-Aided Modelling and Simulation*. Academic Press, Inc., USA.

Stachowiak, H. (1973). *Allgemeine Modelltheorie*. Springer-Verlag, Wien.

Vergne, N. (2008). Drifting Markov models with polynomial drift and applications to DNA sequences. *Statistical Applications in Genetics and Molecular Biology*, 7(1):1–43.

Wenninger, S. and Lames, M. (2016). Performance analysis in table tennis–stochastic simulation by numerical derivation. *International Journal of Computer Science in Sport*, 15(1):22–36.

Zhang, H. (2006). *Leistungsdiagnostik im Tischtennis*. Kovac, Hamburg.

Zhang, H. and Hohmann, A. (2005). Theory and practice of performance diagnosis through mathematical simulation on ball game. *China Sport Science*, 25(8):39–44.

Chapter 8

Statistical Developments in the Quantification of the Inner Game in Tennis

STEPHANIE KOVALCHIK
ZELUS ANALYTICS

Abstract

The mental side of sport continues to be the most elusive dimension of performance analysis. The availability of larger and richer datasets combined with the advancement of machine intelligence is facilitating the first large-scale quantification of aspects of mental performance in elite sport. This chapter summarises these recent advances through the lens of professional tennis. We will begin by contrasting the quantification of mental performance against traditional qualitative methods. This chapter will describe how statistical modelling and machine learning, the two major approaches for quantification in modern data science, are creating novel ways to quantify the "inner game". We will detail how statistical models are being applied to measure game pressure and evaluate its influence on performance. We will also show how deep learning methods are beginning to make breakthroughs in the measurement of emotional expression from video of competitive matches and enable the first direct investigations of the link between player emotionality and performance.

8.1 Introduction

A moment that received much discussion in the first week of the 2018 US Open, one of the four biggest events of the professional tennis season known as the Grand Slams[1], was over an event that took place on the sidelines. Umpires of tennis matches rarely communicate with players during matches without being prompted. However, when Nick Kyrgios went down a break in the second set after already losing the first set, chair umpire Mohamed Lahyani was so discomfited by the mental funk Kyrgios seemed to be in that he got down from his chair and proceeded to give the young Australian a pep talk. With encouraging phrases like "you're great for tennis" and "I know this is not you", Lahyani tried to pull Kyrgios out of his funk. When play resumed, Kyrgios rallied and went on to win the next three sets. Lahyani, on the other hand, was widely censured by the public as well as the ATP Tour who issued him a brief suspension after his interference[2].

It is fitting to begin a discussion of the study of mentality in tennis with an anecdote because so much of widely held views about the link between psychology and performance in sport is based on anecdotal evidence (Sheard, 2012). A close reading of this incident highlights this point and several other issues that are central to the study of mentality in tennis.

Foremost among them is the individual nature of the sport, which requires players to work through the ups and downs of competition on their own, without the aid of a coach or teammate. In trying to motivate Kyrgios, Lahyani violated this fundamental principle of tennis and drew the ire of much of the tennis public. But the structure of the sport was also a key factor to what transpired, as the long breaks between points not only gave Kyrgios the opportunity to brood but also gave Lahyani the opportunity to intervene. During a typical tennis match, players spend 80% of the time in-between points (Harwood, 2016). In other words, players spend fourfold as much time reflecting and reacting to what has happened during competition as they spend actively competing.

Before Lahyani's words, Kyrgios seemed headed for a loss. After the pep talk, he went on to win the next three sets. That reversal made

---

[1] The Grand Slams are the four most prestigious professional events of the tennis season that include the Australian Open, the French Open, the Wimbledon Championships, and the US Open.

[2] https://www.nytimes.com/2018/09/18/sports/tennis/mohamed-lahyani-nick-kyrgios.html

many conclude that Lahyani's words had changed Kyrgios' attitude and that was the change that made his turnaround possible. Implicit in all those conclusions, it is the belief that a player's thoughts and feelings can have a direct influence on their performance. Finally, the widely held views of the public and the sports industry about the link between mentality and performance have significant consequences for players, officials, and other stakeholders in the sport. It was the belief in this link that fuelled the backlash against Lahyani and his ultimate suspension from officiating; and it is the belief in this link that fuels the ongoing scrutiny of every word and gesture made by the mercurial Nick Kyrgios, which has overshadowed the career trajectory of one of the brightest talents to emerge in recent tennis history.

How scientists would attempt to understand what makes a player like Nick Kyrgios tick depends greatly on their field and its preferred research tools. This chapter begins by reviewing the preferred methodologies of cognitive and social scientists–two fields with the longest history of studying the link between athlete psychology and performance–and the major findings about mentality and high performance tennis these approaches have produced. The chapter will then turn its focus to emerging quantitative techniques from the fields of data science and computer vision that are beginning to address the limitations of traditional methods and take the study of mentality in sport into previously unexplored directions.

8.2 Experimental Studies

Controlled experiments are the strongest form of evidence of a presumed cause and its effect. While many questions about an athlete's "mental game" boil down to questions about an action and its result (Did a player's anger cause them to lose a point? Did producing a mental image of the service motion help them execute that motion more effectively?), experimentation is often infeasible for most questions owing to the barriers of reproducing and manipulating psychological states of elite athletes in a lab setting. Several cognitive strategies for skill acquisition have been exceptions to this general rule.

Self-talk is one cognitive strategy that aims to enhance performance through explicit verbalisation of targeted goals. Several experimental studies of high-performance tennis players have investigated the effects of different self-talk strategies on performance through increasing attentional control (Landin and Hebert, 1999) or increasing motivation (Hatzigeorgiadis et al., 2008).

The enhancement of focus and control has also been a central theme of mental imagery or visualisation strategies in tennis. "Visualisation" can refer to internal or external visual cues. Mental imagery is the main mechanism for internal cues, such as a player mentally rehearsing a skill; while external imagery is any visual representation of a player's actions that is from the point-of-view of an external observer, such as in a video image (Dana and Gozalzadeh, 2017; Malouff et al., 2008). Because internal and external visual cues can be controlled by the experimenter, multiple studies have experimentally tested the efficacy of different visualisation approaches for tennis skill acquisition. Ducrocq and colleagues used an experimental design to test short-term effects of a visual search task on inhibitory control of tennis players and presented findings of increased short-term control with greater visualisation training (Ducrocq et al., 2016). When Guillot and colleagues had players complete six weeks of mental imagery training of service skills, they found increased accuracy compared to a control group (Guillot et al., 2013). Dana and Gozalzadeh (2017) conducted a controlled experiment of both internal and external visual cues and their effect on stroke accuracy and found different effectiveness of each type of cue depending on the type of stroke, which suggests that both visualisation types could be important components of skill development (Dana and Gozalzadeh, 2017). In a recent study by Moen et al. a neurotracking task was developed to test the executive functioning of tennis players (Moen et al., 2018). As new computing and neuroscientific tools emerge, we can anticipate that cognitive scientists doing research on sports performance will have the ability to measure more cognitive skills more precisely in experimental settings.

8.3 Qualitative Studies

Whereas cognitive scientists working in sport have been primarily interested in neurological mechanisms of learning and how specific mental skills can enhance the learning process, behavioural scientists and sports psychologists have focused on describing the psychological profile of elite athletes and understanding the role of player psychology on success in high-performance sport. The semi-structured interview has been a mainstay of qualitative research in sports psychology. As defined by Ayers, "The semi-structured interview is a qualitative data collection strategy in which the researcher asks informants a series of predetermined but open-ended questions" (Ayres, 2008). The semi-structured interview allows researchers to "adopt an attitude of curiosity, inviting participants to elaborate on a point, clarify it, and or add more detail in order to fill out the picture of whatever the researcher is trying to understand" (Smith and Caddick, 2012). Whether used in one-on-one interviews or focus groups, the semi-structured interview with its mix of closed- and open-ended questions allows researchers to elicit both focused and more unstructured responses according to the goals of the study.

A large body of qualitative research has focused on how psychological skills and attributes can facilitate athlete development (Abbott and Collins, 2004), transition (Gulbin et al., 2013) and high performance (Weissensteiner et al., 2012). A common conclusion from this research is that athletes who successfully move up to the senior elite ranks display higher levels of motivation, commitment, determination, and drive compared to lower-ranked athletes (Bruner et al., 2008; Holt and Dunn, 2004). Further work to understand the benefits of the psychological skills of self-regulation, motivation, and self-efficacy have argued for their role in goal-setting and attainment, as well as their ability to help athletes manage the stressors that are inherent in the pursuit of a career in sport (Covassin and Pero, 2004; Crespo and Reid, 2007; Gould et al., 1996).

There is one concept in sports psychology that has come to encompass multiple desirable psychological traits in sport and that is "mental toughness". Mental toughness (MT) is the single most studied psychological construct in sport, having hundreds of articles and books published on the topic in the past decade (Sheard, 2012). Interviews with elite athletes and coaches have been a major source of data to understand the qualities of mental toughness, and this information has in turn been used to develop conceptual models for MT (Gucciardi and Gordon, 2013). Although there is no single agreed upon conceptualisation of MT, traits of confidence, drive, resilience, and coping have been some of the most common among multiple proposed definitions over the past two decades (Coulter et al., 2010; Hardy et al., 2014; Thelwell et al., 2005).

While qualitative methods have helped researchers to describe and develop definitions for different aspects of mentality in sport, they are insufficient for measuring and validating these mental skills (Gucciardi et al., 2013). Indeed, the many different definitions of MT proposed by qualitative researchers have caused some to challenge the very concept itself (Caddick and Ryall, 2012). A more common reaction has been to look at the ways in which qualitative research methods have contributed to ambiguities around MT and other complex traits in sports psychology (Gucciardi et al., 2012). Investigators have criticised qualitative work most vociferously for its lack of rigour, reproducibility, and representative samples (Tenenbaum and Driscoll, 2005); and a growing body of quantitative work into mentality in sport has emerged in response to these critiques (Biddle et al., 2001).

8.4 Quantitative Studies

Foremost among quantitative approaches in sports psychology has been the development of structured questionnaires and survey instruments. A manual for MT presented six instruments for this single concept alone (Andersen, 2011): the Psychological Performance Inventory (Loehr, 1986), Psychological Performance Inventory A (Golby et al., 2007), Sport Mental Toughness Questionnaire (Sheard et al., 2009),

Table 8.1: Sample of 9 questions from the 48-item Mental Toughness Questionnaire (Clough et al., 2002)

Question
1. I usually find something to motivate me.
2. I generally feel in control.
3. I generally feel that I am a worthwhile person.
4. Challenges usually bring out the best in me.
5. When working with other people I am usually quite influential/ inspiring.
6. Unexpected changes to my schedule generally throw/bother me.
7. I do not usually give up under pressure.
8. I am generally confident in my own abilities.
9. I usually find myself just going through the motions.
Possible answers to each statement are "strongly disagree", "disagree", "neither agree or disagree", "agree", or "strongly agree".

Australian Football Mental Toughness Inventory (Gucciardi and Gordon, 2009), Cricket Mental Toughness Inventory (Gucciardi and Gordon, 2009), Mental Toughness Questionnaire-48 (MTQ48) (Clough et al., 2002). The goal of these instruments is to provide a tool that can reliably and validly measure the underlying psychological trait in the target athlete population in a way that can support applied studies. The MTQ48, a 48-item instrument developed with responses from 963 mixed student, athlete, and occupational based participants, has become one of the most popular approaches among mental toughness researchers. A sample of questions from the MTQ48 is given in Table 8.1. Meeting these goals, however, is an ongoing challenge of sports psychology research. Gucciardi and colleagues assessed the six instruments designed to measure MT in terms of their 'conceptual, statistical and empirical' grounds and concluded that "no measure sufficiently satisfies all three criteria" (Gucciardi et al., 2013), suggesting that there is currently no single preferred instrument for this most-discussed psychological trait.

Despite uncertainties over the optimal survey measurement of mental traits of elite athletes, numerous studies have examined the association between scores on available psychometric instruments and performance in sport. A majority of studies reviewed by Cowden found that mentally tougher athletes participated in higher levels of competition and showed superior performance in competition compared to athletes scoring lower (Cowden, 2016; Cowden et al., 2014). These findings are in line with broader evidence of a correlation between MT inventory scores and perseverance and motivation across

8.4. QUANTITATIVE STUDIES

multiple sports (Gucciardi et al., 2015, 2016).

The development of inventories and questionnaires in the study of mentality and psychological traits in sport has addressed some of the limitations of qualitative approaches while leaving others unresolved. Questionnaire instruments improve on semi-structured interviews by allowing for psychometric evaluation, which can test the instrument against well-defined statistical standards. The availability of instruments also helps to improve consistency in data collection across applications in different samples and conducted by different researchers. Despite these strengths, these traditional quantitative approaches still have a fundamental reliance on retrospective self-report. A clear limitation of this reliance is that it prohibits measurement during competition, the time that is of primary interest for most questions about mentality and its effect on performance. Further, in employing self-report inventories, there is a basic supposition that it is possible to learn about the psychological mechanism of athletes through their own reflections and perception, a premise that is neither certain nor verifiable (Saw et al., 2015).

8.4.1 Emerging Tools of Quantitative Analysis

The labour-intensive nature of inventory development and the inherent shortcomings of even the best designed instruments have prompted some researchers to consider alternative methods to analyse the mental aspects of elite tennis performance. Owing to the lack of in-competition data with traditional methods, there is an increasing interest among sports researchers to use in-competition performance data to investigate the psychological skills and decision-making in elite tennis. In recent years, the pursuit of this interest has resulted in two major directions of quantitative methodology: indirect measurement of unexplained variation in performance using regression modelling and direct measurement of emotional expression in competition using computer vision and machine learning. The remainder of this section summarises developments in each of these lines of research.

With the increasing richness of available match data, researchers are beginning to explore what can be learned about the mental side of the game from the statistical analysis of variation in in-match performance. Clutch performance is a popular psychological concept in sport that refers to an athlete's ability to perform effectively in situations of high pressure where the outcome of the current moment could have a high impact on the game outcome. Statistical measures of clutch performance in tennis have included above average win percentages at Grand Slams (Jetter and Walker, 2015) and above average performance on break points in a match (González-Díaz et al., 2012). Kovalchik and Reid developed a general methodology for quantifying performance under pressure using a method they call "clutch averaging", which weights point-level performance by a leverage weight representing the possible shift in win expectations based on the outcome of the current point (Kovalchik and Reid, 2017).

Another aspect of in-competition mentality that has been explored in several tennis studies is momentum. In a sport context momentum describes an improvement in performance when a player has had a successful run in a competition, a phenomenon that is sometimes referred to as "streakiness" or a "hot hand" effect. Moss and O'Donoghue investigated momentum effects at the point-level using a sample of matches from the 2013 US Open Men's Championship (Moss and O'Donoghue, 2015). The authors found that the outcome of the previous point or previous three points had no significant influence on performance on the current point. By contrast, Klaassen and Magnus investigated the same question using a larger set of matches from the 1992-1995 Wimbledon Championships, and they did find significant momentum effects (Klaassen and Magnus, 2001). Moreover, Klaassen and Magnus found that both the momentum and point importance varied with player rankings in a direction suggesting that better players were characterised by greater consistency in performance across game situations.

8.4.2 Identifying Mentality Profiles

Whereas previous studies of variation in performance have primarily focused on changes in response to a single type of pressure situation in tennis, a recent paper by Kovalchik and Ingram brought together multiple game situations to develop multi-dimensional mental profiles for tennis players (Kovalchik and Ingram, 2016). Analysing 3 million points played in professional tennis matches, these authors applied hierarchical clustering to the situational effects of players on serve and return for players competing in Grand Slam events in 2015. The situational effects were based on the estimates of how player win percentages on points changed with eleven contextual factors: in tiebreaks, when on a break point, when a point away from a break point, when a set up in the match, when a set down in the match, on more important points, with changes in point spread, when the last point was won, when a service break chance was missed, when a service break chance was saved, and with game length. After standardisation each player was represented by their effect vector and the hierarchical clustering was applied to the distance measures between all pairs of players, grouping players according to the similarity in their distance values against the rest of the sample (Anderberg, 2014).

From this analysis they identified eight mentality profiles for male and female players, each profile corresponding to a different pattern of response to situational pressures. A summary of the male profiles and the game situations considered is shown in Figure 8.1. Each line in a plot denotes a specific player's set of dynamic effects on serve and return, with effects scaled to have an equal standard deviation of one.

8.4. QUANTITATIVE STUDIES

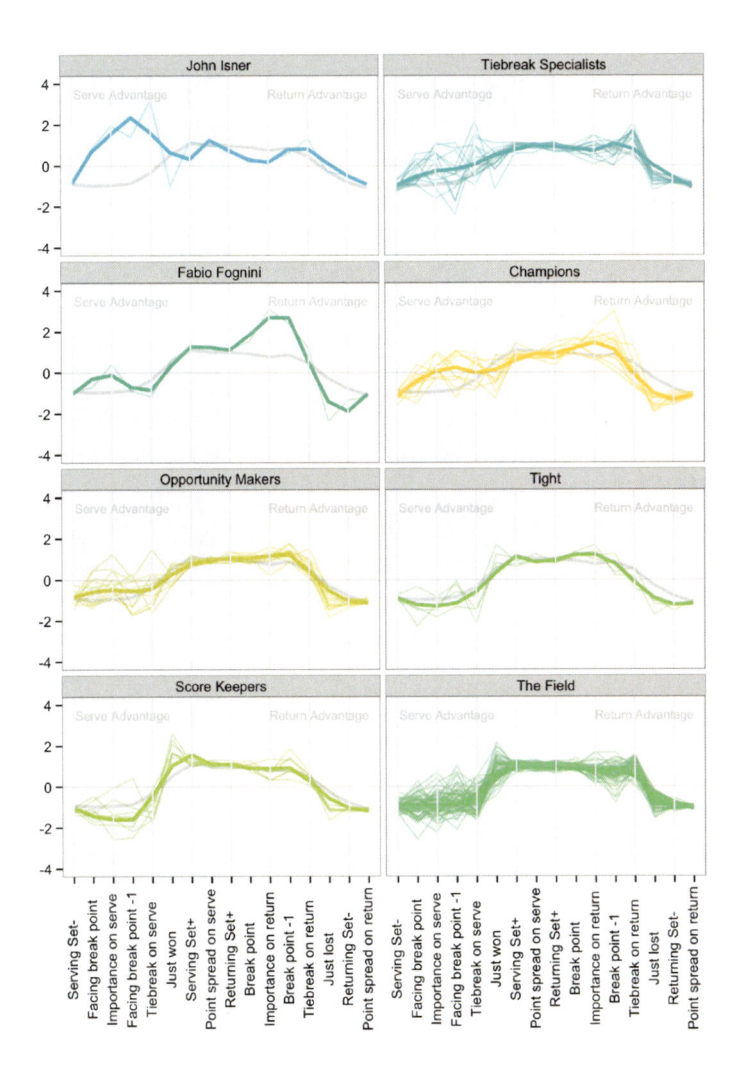

Figure 8.1: Reproduction of parallel coordinates plot of mentality profiles of professional men's tennis players presented in (Kovalchik and Ingram, 2016)

A smoothed regression line is plotted over the observed profiles in each panel to highlight the key differences from the status quo ("The Field") shown in grey.

To convey some of the differences among these profiles, we summarise three of the profiles found. One of two players with a unique profile was big server John Isner, who had the largest positive effects across conditions on serve; the other player with their own profile was Fabio Fognini, known for his mercurial temperament, who showed mentally strong effects on more important points (especially on the return game) and on making break point opportunities, yet large negative effects when a point or set down on return. Among active players who currently hold the most Grand Slam titles—Novak Djokovic, Roger Federer, Andy Murray and Rafael Nadal (colloquially referred to as the "Big Four")—all shared a common profile. These players were distinctive in exhibiting the service strength of tiebreak specialists but also high mental strength on the return when playing more important points. Based on this finding, the authors described this profile as a "Champion's mentality".

Psychological effects are one possible cause of the observed variation in performance identified in the above examples, but they are not the only possible cause. Strategic adaptation, for instance, could be an equally plausible explanation for situation-specific changes in performance. While there is inherent value in identifying and measuring systematic variation in performance, the ambiguity in attributing the cause of performance variation limits the potential impact of indirect effects for progressing sports psychology.

8.4.3 Emotion Measurement with Computer Vision

Advances in imaging processing and data capture from broadcast feeds are creating opportunities for the direct measurement of the emotionality of athletes in competition. The ability to detect basic emotions and facial expressions from static images using machine learning techniques is now well-established (Calvo and D'Mello, 2010). The first extension of these methods to tennis was presented by Kovalchik and Reid who developed a modelling framework (see Figure 8.2) for predicting the occurrence and intensity of seven "sport-relevant" emotions of professional tennis athletes using a single-camera video (Kovalchik and Reid, 2018). The model training was based on a one-of-a-kind dataset of 1,700 facial images of tennis players labelled with the sport-relevant emotions on a 10-point scale by five raters per emotion and image. These emotion labels were the outcomes, while the explanatory variables were facial action units describing micromovements of the expression that were detected with established deep learning techniques (Baltrusaitis et al., 2018). The authors then compared the predictive performance of 15 different machine learning techniques and found that classification with support vector machines provided good out-of-sample performance across the emotion

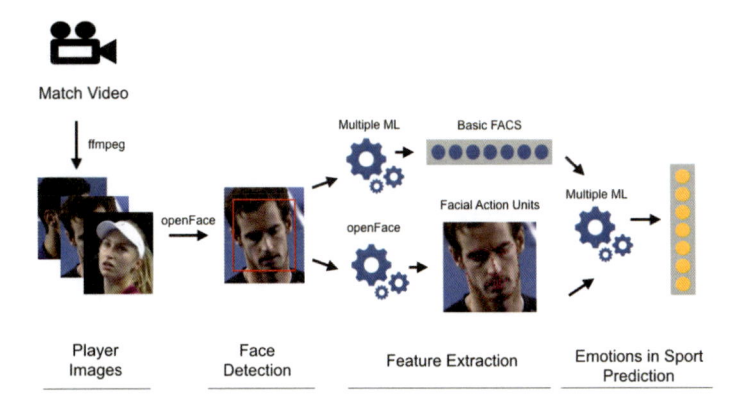

Figure 8.2: Reproduction of real-time emotion capture framework for tennis broadcasts presented in (Kovalchik and Reid, 2018). The purpose of the framework is to use image frames from broadcast video, capture facial expression using computer vision and machine learning (ML) techniques, and use ML classification methods to predict the emotion conveyed by the captured expression.

categories.

The authors used a similar data collection strategy to obtain thousands of in-competition images for the "Big Four" players (as referenced above). Once the facial action units and prediction emotions were obtained, data was also tabulated regarding the score outcomes of the match based on the time each image occurred. This enables the authors to assess how elite tennis players emotionally reacted to point outcomes and how reactions predicted future point outcomes. The resulting emotional profiles not only established the face validity of the method but also provided the first evidence of a direct link between in-competition performance outcomes and emotions of professional tennis players.

8.5 Summary

The "inner game" of sport has long been one of the most captivating topics in the pop culture of sport, yet one of the most intractable areas of study for sports scientists. As an individual sport, tennis provides a fitting microcosm of the broader methodological trends in the study of player psychology and its link to performance. This chapter has provided an overview of several decades of qualitative and quantitative research in tennis that have used interviews and structured inventories to begin to describe, conceptualise, and operationalise previously elusive concepts like "mental toughness". While these conventional methods have established some evidence on athlete perceptions about their psychology and its role in athlete development, skill acquisition, and performance, researchers have also begun to explore approaches for quantifying the psychological traits of elite athletes in competition without the dependency of self-report. It is no coincidence that these advances have coincided with an explosion in tracking data, wearable technology, and computer vision in sport. As new technologies continue to make richer and more granular in-competition data available to researchers, it can be expected that more data-centred, model-driven approaches will become the future of quantification of the mental game in sport performance.

Bibliography

Abbott, A. and Collins, D. (2004). Eliminating the dichotomy between theory and practice in talent identification and development: considering the role of psychology. *Journal of Sports Sciences*, 22(5):395–408.

Anderberg, M. R. (2014). *Cluster Analysis for Applications*. Academic Press, New York.

Andersen, M. B. (2011). Who's mental, who's tough and who's both? Mutton constructs dressed up as lamb. In Gucciardi, D. F. and Gordan, S., editors, *Mental Toughness in Sport: Developments in Theory and Research*, pages 69–88. Routledge, East Sussex, UK.

Ayres, L. (2008). Semi-structured interview. In *The SAGE Encyclopedia of Qualitative Research Methods*, pages 811–812. SAGE Publications, Thousand Oaks, California.

Baltrusaitis, T., Zadeh, A., Lim, Y. C., and Morency, L. (2018). Openface 2.0: Facial behavior analysis toolkit. In *2018 13th IEEE International Conference on Automatic Face & Gesture Recognition (FG 2018)*, pages 59–66. IEEE Computer Society.

Biddle, S. J. H., Markland, D., Gilbourne, D., Chatzisarantis, N. L. D., and Sparkes, A. C. (2001). Research methods in sport and exercise psychology: Quantitative and qualitative issues. *Journal of Sports Sciences*, 19(10):777–809.

Bruner, M. W., Munroe-Chandler, K. J., and Spink, K. S. (2008). Entry into elite sport: A preliminary investigation into the transition experiences of rookie athletes. *Journal of Applied Sport Psychology*, 20(2):236–252.

Caddick, N. and Ryall, E. (2012). The social construction of 'mental toughness'–a fascistoid ideology? *Journal of the Philosophy of Sport*, 39(1):137–154.

Calvo, R. A. and D'Mello, S. (2010). Affect detection: An interdisciplinary review of models, methods, and their applications. *IEEE Transactions on Affective Computing*, 1(1):18–37.

Clough, P., Earle, K., and Sewell, D. (2002). Mental toughness: The concept and its measurement. In *Solutions in Sport Psychology*, pages 32–43.

Coulter, T. J., Mallett, C. J., and Gucciardi, D. F. (2010). Understanding mental toughness in Australian soccer: Perceptions of players, parents, and coaches. *Journal of Sports Sciences*, 28(7):699–716.

Covassin, T. and Pero, S. (2004). The relationship between self-confidence, mood state, and anxiety among collegiate tennis players. *Journal of Sport Behavior*, 27(3):230–242.

Cowden, R. G. (2016). Competitive performance correlates of mental toughness in tennis: A preliminary analysis. *Perceptual and Motor Skills*, 123(1):341–360.

Cowden, R. G., Fuller, D. K., and Anshel, M. H. (2014). Psychological predictors of mental toughness in elite tennis: An exploratory study in learned resourcefulness and competitive trait anxiety. *Perceptual and Motor Skills*, 119(3):661–678.

Crespo, M. and Reid, M. M. (2007). Motivation in tennis. *British Journal of Sports Medicine*, 41(11):769–772.

Dana, A. and Gozalzadeh, E. (2017). Internal and external imagery effects on tennis skills among novices. *Perceptual and Motor Skills*, 124(5):1022–1043.

Ducrocq, E., Wilson, M., Vine, S., and Derakshan, N. (2016). Training attentional control improves cognitive and motor task performance. *Journal of Sport and Exercise Psychology*, 38(5):521–533.

Golby, J., Sheard, M., and Van Wersch, A. (2007). Evaluating the factor structure of the psychological performance inventory. *Perceptual and Motor Skills*, 105(1):309–325.

González-Díaz, J., Gossner, O., and Rogers, B. W. (2012). Performing best when it matters most: Evidence from professional tennis. *Journal of Economic Behavior & Organization*, 84(3):767–781.

Gould, D., Tuffey, S., Udry, E., and Loehr, J. (1996). Burnout in competitive junior tennis players: I. A quantitative psychological assessment. *The Sport Psychologist*, 10(4):322–340.

Gucciardi, D. and Gordon, S. (2013). Mental toughness in sport: past, present and future. In Gucciardi, D. F. and Gordan, S., editors, *Mental Toughness in Sport: Developments in Theory and Research*, pages 233–251. Routledge, East Sussex, UK.

Gucciardi, D. F. and Gordon, S. (2009). Development and preliminary validation of the cricket mental toughness inventory (cmti). *Journal of Sports Sciences*, 27(12):1293–1310.

Gucciardi, D. F., Hanton, S., and Mallett, C. J. (2012). Progressing measurement in mental toughness: A case example of the mental toughness questionnaire 48. *Sport, Exercise, and Performance Psychology*, 1(3):194.

Gucciardi, D. F., Jackson, B., Hanton, S., and Reid, M. (2015). Motivational correlates of mentally tough behaviours in tennis. *Journal of Science and Medicine in Sport*, 18(1):67–71.

Gucciardi, D. F., Mallett, C. J., Hanrahan, S. J., and Gordon, S. (2013). Measuring mental toughness in sport. In Gucciardi, D. F. and Gordan, S., editors, *Mental Toughness in Sport: Developments in Theory and Research*, pages 108–32. Routledge, East Sussex, UK.

Gucciardi, D. F., Peeling, P., Ducker, K. J., and Dawson, B. (2016). When the going gets tough: Mental toughness and its relationship with behavioural perseverance. *Journal of Science and Medicine in Sport*, 19(1):81–86.

Guillot, A., Desliens, S., Rouyer, C., and Rogowski, I. (2013). Motor imagery and tennis serve performance: The external focus efficacy. *Journal of Sports Science & Medicine*, 12(2):332–338.

Gulbin, J., Weissensteiner, J., Oldenziel, K., and Gagné, F. (2013). Patterns of performance development in elite athletes. *European Journal of Sport Science*, 13(6):605–614.

Hardy, L., Bell, J., and Beattie, S. (2014). A neuropsychological model of mentally tough behavior. *Journal of Personality*, 82(1):69–81.

Harwood, C. (2016). Twenty years' experience working within professional tennis. In *Psychology in Professional Sports and the Performing Arts*, pages 91–106. Routledge.

Hatzigeorgiadis, A., Zourbanos, N., Goltsios, C., and Theodorakis, Y. (2008). Investigating the functions of self-talk: The effects of motivational self-talk on self-efficacy and performance in young tennis players. *The Sport Psychologist*, 22(4):458–471.

Holt, N. L. and Dunn, J. G. H. (2004). Toward a grounded theory of the psychosocial competencies and environmental conditions associated with soccer success. *Journal of Applied Sport Psychology*, 16(3):199–219.

Jetter, M. and Walker, J. K. (2015). Game, set, and match: Do women and men perform differently in competitive situations? *Journal of Economic Behavior & Organization*, 119:96–108.

Klaassen, F. J. G. M. and Magnus, J. R. (2001). Are points in tennis independent and identically distributed? Evidence from a dynamic binary panel data model. *Journal of the American Statistical Association*, 96(454):500–509.

Kovalchik, S. and Ingram, M. (2016). Hot heads, cool heads, and tacticians: Measuring the mental game in tennis (ID: 1464). In *MIT Sloan Sports Analytics Conference*.

Kovalchik, S. and Reid, M. (2017). Measuring clutch performance in professional tennis. *Italian Journal of Applied Statistics*, 30(2):255–268.

Kovalchik, S. and Reid, M. (2018). Going inside the inner game: Predicting the emotions of professional tennis players from match broadcasts. In *MIT Sloan Sports Analytics Conference*.

Landin, D. and Hebert, E. P. (1999). The influence of self-talk on the performance of skilled female tennis players. *Journal of Applied Sport Psychology*, 11(2):263–282.

Loehr, J. (1986). *Mental Toughness Training for Sports: Achieving Athletic Excellence*. Stephen Greene Press, Lexington, Massachusetts.

Malouff, M. J., McGee, J., Halford, T. H., and Rooke, E. S. (2008). Effects of pre-competition positive imagery and self-instructions on accuracy of serving in tennis. *Journal of Sport Behavior*, 31(3):264–275.

Moen, F., Hrozanova, M., and Stiles, T. (2018). The effects of perceptual-cognitive training with neurotracker on executive brain functions among elite athletes. *Cogent Psychology*, 5(1):1544105.

Moss, B. and O'Donoghue, P. (2015). Momentum in US Open men's singles tennis. *International Journal of Performance Analysis in Sport*, 15(3):884–896.

Saw, A. E., Main, L. C., and Gastin, P. B. (2015). Monitoring athletes through self-report: factors influencing implementation. *Journal of Sports Science & Medicine*, 14(1):137–146.

Sheard, M. (2012). *Mental Toughness: The Mindset Behind Sporting Achievement*. Routledge.

Sheard, M., Golby, J., and Van Wersch, P. (2009). Progress toward construct validation of the sports mental toughness questionnaire (smtq). *European Journal of Psychological Assessment*, 25(3):186–193.

Smith, B. and Caddick, N. (2012). Qualitative methods in sport: A concise overview for guiding social scientific sport research. *Asia Pacific Journal of Sport and Social Science*, 1(1):60–73.

Tenenbaum, G. and Driscoll, M. P. (2005). *Methods of Research in Sport Sciences: Quantitative and Qualitative Approaches*. Meyer & Meyer Verlag.

Thelwell, R., Weston, N., and Greenlees, I. (2005). Defining and understanding mental toughness within soccer. *Journal of Applied Sport Psychology*, 17(4):326–332.

Weissensteiner, J. R., Abernethy, B., Farrow, D., and Gross, J. (2012). Distinguishing psychological characteristics of expert cricket batsmen. *Journal of Science and Medicine in Sport*, 15(1):74–79.

Chapter 9

Informational Content of Tennis Betting Odds

RUUD H. KONING
UNIVERSITY OF GRONINGEN

TOM BOOT
UNIVERSITY OF GRONINGEN

Abstract

One of the defining characteristics of sports matches is that the outcome of the match is uncertain when it is started. Even an overwhelmingly favourite team or athlete can lose against an underdog. This uncertainty of outcome is one of the scarce commodities produced by (professional) sporting contests (Fort, 2011). Another scarce good produced by a sporting contest is commonality. People discuss a sporting contest before it is actually played, and this discussion is frequently about the likely winner. One more or less objective measure of the likely outcome of a sport match are betting odds: payout on an event is high if it is unlikely, and low if it is likely. In this chapter we focus on the informational content of betting odds, in the context of professional tennis. Are betting odds a good predictor of the outcome of a tennis match, or are other covariates relevant as well? Examples of potentially relevant covariates are type of tournament, history of the player, surface of the tournament, etc. Also, our dataset contains betting odds from different betting firms, so we also assess whether informational content is lower if there is more disagreement about the betting odds for a particular match between those firms. The analysis in this chapter has broader relevance. If we can show that this particular betting market processes information efficiently, we can be more confident that similar markets defined on other outcomes are also informationally efficient. In fact, informational efficiency of betting markets may be a reason to open markets on special events. Pooling the information of all participants may provide a good indication of the likelihood of the outcome.

9.1 Introduction

One of the defining characteristics of sports matches is that the outcome of the match is uncertain when it is started. Even an overwhelmingly favourite team or athlete can lose against an underdog. This uncertainty of outcome is one of the scarce commodities produced by (professional) sporting contests (Fort, 2011). Another scarce good produced by a sporting contest is commonality. People discuss a sporting contest before it is actually played, and this discussion is frequently about the likely winner. One more or less objective measure of the likelihood of a certain outcome in a sport match is betting odds: pay-out on a possible outcome is high if it is unlikely and low if it is likely.

As an example, consider the 2018 Wimbledon men's final between Anderson and Djokovic. The pay-out on Anderson to win was 5, and the pay-out on Djokovic to win was 1.16. We will argue that these decimal odds translate into implied winning "probabilities" of 0.20 ($= 1/5$) and 0.86 ($= 1/1.16$) respectively, indicating the likelihood of either player to win. These numbers add up to more than 1, 1.06 in this case, which is called the booksum. The amount that 1 is exceeded is known as the overround (here, 0.06). The main focus of this chapter is to transform the implied winning "probabilities" that add up to the booksum into proper probabilities that add up to 1, and we assess how these proper probabilities are unbiased estimators of the outcomes of tennis matches.

The suggestion to use betting odds as indicators of likely match outcomes immediately raises the question: how well do betting odds predict the outcome of a tennis match? In other words, what is the informational content of tennis betting odds? In this chapter, we answer this question based on a large dataset containing most important tennis tournaments between 2001 and 2018 for both men and women.

We find that unbiased estimates of winning probabilities can be inferred directly from betting odds through the model of Shin (1993). This finding is robust to using grouped level or individual match data. The only exception we find is when we specialise to Grand Slam matches, where a favourite-longshot bias appears.

The analysis in this chapter has broader relevance. If we can show that this particular betting market processes information efficiently, we can be more confident that similar markets defined on other outcomes are also informationally efficient. In fact, informational efficiency of betting markets may be a reason to open markets on special events. Pooling the information of all participants may provide a good indication of the likelihood of the outcome.

The setup of this chapter is as follows. We start with a discussion of some relevant literature in Section 9.2. Some betting terminology

9.2 Literature Review

Various concepts have been developed to define and measure betting market efficiency. A betting market is said to be weakly efficient if the expected return on all bets is the same. A betting market is semi-strong efficient if, based on public information, it is not possible to devise a betting strategy that earns abnormal returns. The strong form of efficiency extends the information set to include private information. Reviews on the literature on betting market efficiency are given by Williams (1999), Williams (2005), Sauer (1998), and Sauer (2005).

Weak form efficiency is at odds with the existence of a favourite-longshot bias, in which underdogs are overvalued and favourites undervalued. The existence of this bias has been well-documented for horse race betting, see Williams (1999) for an overview. Theoretically, the favourite-longshot bias can be explained from supply-side (bookmakers) arguments as well as demand-side (bettors) arguments. The most common explanation from the perspective of the demand side is that bettors are risk-loving (Quandt, 1986). However, this is not the only explanation. Golec and Tamarkin (1995) show that the bias can arise from overconfidence of bettors, while Golec and Tamarkin (1998) argue that bettors love skewness rather than risk. In men's professional tennis, Forrest and McHale (2007) find that there is evidence both for risk-loving and skewness-loving preferences, that is, a preference for larger wins with smaller probabilities.

A supply side argument for the existence of the favourite-longshot bias is provided by Shin (1993). He shows that it can arise from bookmakers facing a market where a fraction of the bettors has inside knowledge. We discuss this model in some detail in Section 9.3. Shin's model implies a relation between the number of outcomes and the booksum, which is verified empirically by Williams and Paton (1997), lending some support to this supply-side explanation. Schnytzer et al. (2010) also document a substantial amount of inside trading in a horse betting market in Australia. Lahvička (2014) studies the causes of the longshot bias in tennis markets in more detail. Among others, it is found that this bias is stronger for high-profile tournaments. This finding is corroborated by Abinzano et al. (2016), and we come back to it in our empirical analysis.

Although weak form inefficiency of betting markets has been well documented across several sports, there is substantial discussion on the stronger forms of efficiency. In football, Goddard and Asimakopoulos (2004) find some evidence for a betting strategy that appears to yield positive returns for English league matches. The same conclusion is reached for European football matches by Vlastakis et al. (2009). Similar studies on the National Football League (NFL) offer mixed results. Pankoff (1968) cannot find a profitable betting strategy, while Zuber et al. (1985) do report that such strategies exist. Then, on more recent data, Boulier et al. (2006) do not find evidence for market inefficiencies. Stekler et al. (2010) provide an extensive review to compare prediction in different sports (horse racing, baseball, American football, basketball, football). They find that neither forecasts from statistical models, nor from experts, consistently outperform the market.

There are several publicly available factors that might influence the outcome of the match. For tennis matches, Sunde (2009) shows that heterogeneous contestant pools reduce exerted effort by professional tennis players. Koning (2011) shows that home advantage positively affects match outcomes for male tennis players, while female players appear unaffected. Del Corral and Prieto-Rodríguez (2010) find that difference in men's and women's world rankings is significant in determining the outcome of a tennis match. Also, age difference matters, but this effect differs between sexes. The question is whether the above effects are reflected in the betting odds. McHale and Morton (2011) use a Bradley-Terry type model to forecast tennis matches and report positive returns, thus providing evidence against semi-strong market efficiency. There is also an active betting market during the time at which a match takes place. Further research is needed to determine whether in-game betting is profitable. Here, in-game predictions such as those provided by Klaassen and Magnus (2003) could provide valuable information.

A profit maximising (or loss minimising) bettor needs to know the probabilities associated with the different match outcomes. If the quoted odds reflect all available information, one might want to use these odds to infer the true winning probabilities. However, there are multiple ways to translate published odds into probabilities. Inverting the odds, which gives the "implied probability" quoted by various online calculators, leads to probabilities that add up to more than 1. To correct for this, one can divide by the sum of the inverted odds, see for example Forrest et al. (2005). We refer to this method as scaling, it is also called basic normalisation by Štrumbelj (2014) or the multiplicative method by Clarke et al. (2017). Štrumbelj (2014) finds that scaling does not yield an unbiased estimate of the true winning probabilities. Interestingly, he finds that using the model of Shin (1993) to infer probabilities from odds leads to much better results. This is confirmed by Clarke et al. (2017), who additionally compare the scaling method and the Shin probabilities to the power method by

Vovk and Zhdanov (2009). They find that the power method yields comparable results to the use of Shin probabilities in approximating the true winning frequencies.

9.3 Betting Terminology, Notation and Theoretical Background

In this chapter we consider fixed odds betting markets. A bookmaker offers a pay-out on an uncertain event. We assume that these pay-outs are quoted as decimal odds, that is, they give the amount to be paid by the bookmaker if the event occurs. This amount includes the stake by the bettor. Odds are fixed when the bettor and the bookmaker engage in the (implicit) contract, even if the bookmaker decides to adjust the odds later (but before the match).

In a tennis match between players A and B, we denote such fixed odds by O_A and O_B. If A wins, the bettor receives O_A, and his profit is $O_A - 1$. Odds exceed 1, small (short) odds (close to 1) indicate a likely winner, large (long) odds indicate an underdog.

A bet is called fair if its expected profit is 0, which should occur under the assumption of a competitive betting market (multiple bookmakers offer bets on the same tennis match). This allows us to calculate win "probabilities" implicit in the bookmaker's odds. Let $\tilde{p}_A$ be the "probability" that A wins, then a fair bet implies $\tilde{p}_A O_A + (1 - \tilde{p}_A) \times 0 = 1$, so $\tilde{p}_A = 1/O_A$. Similarly, we have for player B that $\tilde{p}_B = 1/O_B$. In practice, these implied "probabilities" add up to more than 1: $\tilde{p}_A + \tilde{p}_B = \tilde{p} \equiv 1 + \lambda$ (which is the reason why "probabilities" has been in quotation marks so far). $\tilde{p}$ is known as the booksum, and λ is known as the overround. The implied "probabilities" are also known as prices.

As mentioned in Section 9.2, an empirical feature documented in different studies related to betting and sport results (especially horse betting) is the favourite-longshot bias: betting on an unlikely winner (a longshot) yields a worse return than betting on likely winners (favourites). For tennis, Forrest and McHale (2007) show that "...underdogs tend to become increasingly unfair betting propositions" and that "the data indicate a positive longshot bias" (p. 763).

There are different ways to distribute the overround λ over the implied "probabilities" $\tilde{p}_A$ and $\tilde{p}_B$, see for example Clarke et al. (2017). We discuss the most common ones.

Scaling A simple approach to obtain probabilities that add up to 1 is to scale the implied "probabilities" by their sum

$$p_A = \frac{\tilde{p}_A}{\tilde{p}}, \quad p_B = \frac{\tilde{p}_B}{\tilde{p}}.$$

This is referred to as the basic normalisation by Štrumbelj (2014) and as the multiplicative method by Clarke et al. (2017). By construction, using scaled probabilities to infer expected profit yields identical profits for bets on player A or player B.

Shin Probabilities An alternative to scaling can be derived from the model of Shin (1993), describing a betting market for horse races. The model explains the favourite-longshot bias that is observed in the data by introducing a fraction of insider traders to the market. These insider traders have perfect knowledge on the outcome of the match. Bookmakers seek insurance against the possibility of facing an insider, and this insurance drives a wedge between relative prices and relative outcome probabilities.

Shin's model has previously been used by Smith et al. (2009) and Štrumbelj (2014) to infer winning probabilities. We provide a short overview of the model and some of its properties here. The market consists of bettors, of which a fraction z are insider traders. The insider traders know the outcome of the tennis match *ex ante*. The bookmaker believes player A wins the match with probability p_A. Given the fraction of insiders z and the probability p_A, Shin's model implies the price for betting on player A or player B.

Figure 9.1 plots the bookmakers prices implied by the model as a function of the true winning probabilities[1]. A notable feature of the implied prices is that when the fraction of insiders is nonzero and the probability that player A wins is high, the bookmakers' insurance against insider knowledge leads to a price slightly larger than 1. In practice, this means that the bookmaker would refuse bets on this particular match. When the fraction of insiders is 0.07, which is a typical fraction implied by our data, this would occur for winning probabilities $p_A > 0.95$.

The above discussion calculates the implied prices from the underlying winning probabilities and the fraction of insiders. Shin's model can also be used in reverse to extract the beliefs of the betting agencies and the implied fraction of insiders from observed betting odds. Clarke et al. (2017) show that in this model true beliefs about the outcome of the tennis match, and the implied "probabilities" are related

---

[1]Equation (5) in Shin (1993) shows that when bookmakers maximise profits, given a fraction of insider traders z and true winning probabilities p_A and p_B, the bookmaker sets the price of a ticket for player A at

$$\tilde{p}_A = zp_A + (1-z)p_A^2 + \sqrt{\left[zp_A + (1-z)p_A^2\right] \cdot \left[zp_B + (1-z)p_B^2\right]}.$$

9.3. BETTING TERMINOLOGY 195

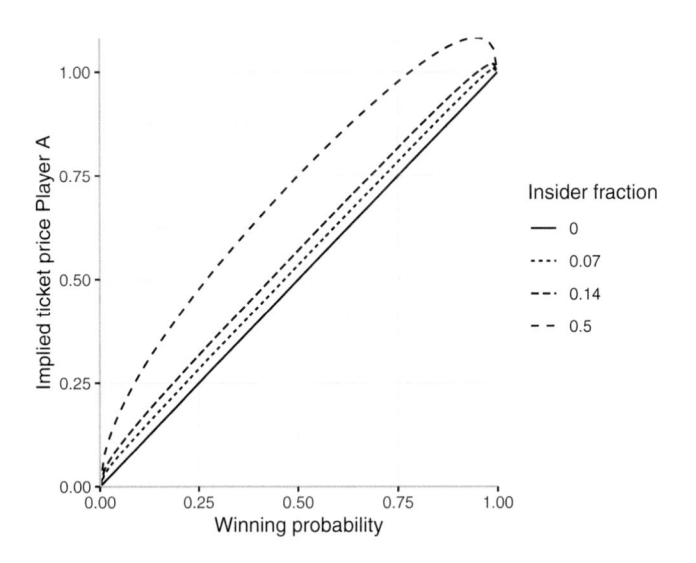

Figure 9.1: Shin (1993): implied price as a function of the true winning probability.

by

$$p_A = \tilde{p}_A - \lambda/2, \quad p_B = \tilde{p}_B - \lambda/2. \tag{9.1}$$

with λ the overround, i.e. $\tilde{p}_A + \tilde{p}_B - 1$. Although in Shin's model the probabilities p_A and p_B are necessarily between zero and one, (9.1) does not prevent negative estimates of the bookmakers' beliefs based on observational data. We, however, do not observe this in our dataset.

We close this discussion by relating two implications of Shin's model to our dataset. For a given insider fraction z and price $\tilde{p}_A$, Shin's model implies the booksum we should find. In our analysis below we estimate the insider fraction for each match separately. The average insider fraction in the data is 0.07. If we make the simplifying assumption that the insider fraction is 0.07 for all matches, the solid line in Figure 9.2 shows the theoretical value of the booksum as a function of the price for a ticket for player A. The dots represent the values of the booksum found in the data as a function of the price (inverse odds) for a ticket for player A. We see that the line captures the general trend reasonably well.

The second implication of Shin's model is that the maximum overround is achieved when $p_A = \frac{1}{2}$, in which case $\tilde{p}_A + \tilde{p}_B = 1 + z$. In

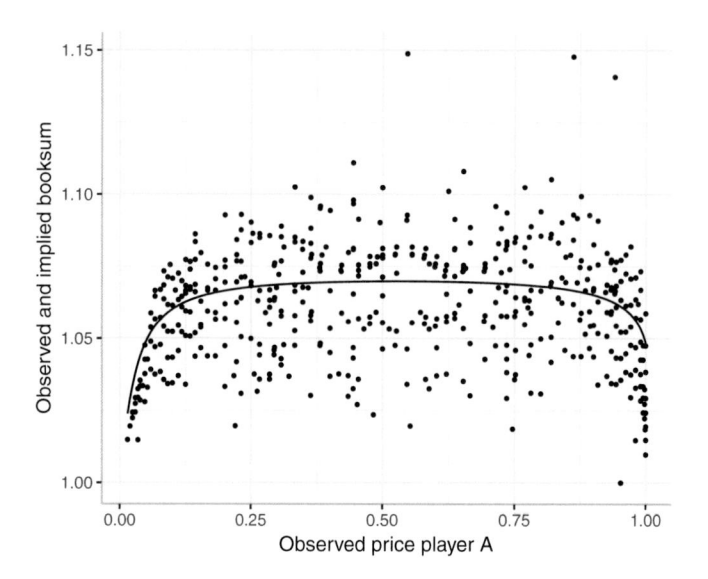

Figure 9.2: Shin (1993): model implied (insider fraction $z = 0.07$) and empirical booksum as a function of the observed price.

Figure 9.2, the highest dot is the match between Marion Bartoli (11) and Samantha Stosur (6) in the finals of the HP Open 2011 in Osaka. Betting odds where 1.83 for Bartoli and 1.66 for Stosur. This match was peculiar in the sense that the semi-finals were played on the same day. Stosur's semi-final lasted more than 2.5 hours, while Bartoli finished in an hour less. Bartoli won the final 6-3 6-1.

Power Method Instead of using a multiplication scaling method or Shin's additive method, a third option is the power method, see for example Clarke et al. (2017). Here, the true winning probabilities are related to the ticket prices as

$$p_A = (\tilde{p}_A)^k, \quad p_B = (\tilde{p}_B)^k.$$

The factor k is found by solving numerically

$$(\tilde{p}_A)^k + (\tilde{p}_B)^k - 1 = 0,$$

ensuring that the probabilities add up to one. The power method also corrects for the favourite-longshot bias, but differently from the Shin probabilities, as we discuss below.

9.3. BETTING TERMINOLOGY

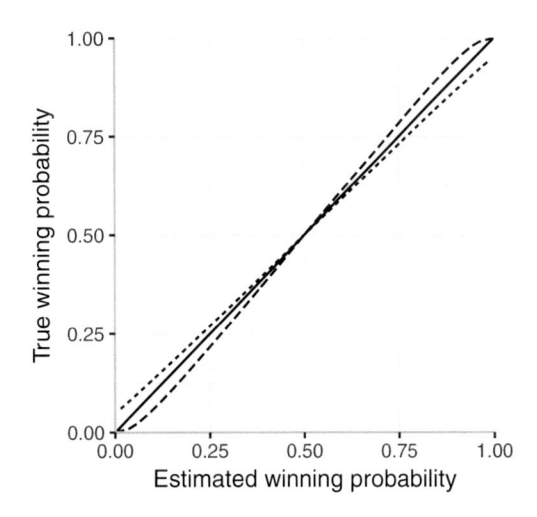

Figure 9.3: True probabilities versus estimated probabilities: projected behaviour under the Shin model (insider fraction $z = 0.14$, long-dashed line: scaled probabilities, short-dashed line: power method probabilities).

Comparison Suppose that the Shin model is indeed the correct model underlying the data. How would the winning probabilities inferred from data on the ticket prices (inverse decimal odds), using the scaling and power method, compare to the true winning probabilities?

Assume that the insider fraction is equal to $z = 0.14$ (this fraction is chosen such that the lines in Figure 9.3 are different enough). For winning probabilities $p_A = \{0, 0.01, 0.02, \ldots, 1\}$, we generate the corresponding prices $\tilde{p}_A$ and $\tilde{p}_B$. Using these implied prices, we then calculate the winning probabilities using the scaling method, the Shin probabilities, and the power method. Of course, since the Shin model is correct, the Shin probabilities correspond one-to-one with the true probabilities.

Figure 9.3 shows that using the scaling method (long-dashed lines) we would observe the favourite-longshot bias: the winning probabilities assigned to favourites are too low, while the probabilities assigned to underdogs are too high. If we were to calculate a linear regression line based on such data, the slope coefficient would be estimated to be larger than one, and the intercept would be negative. Using the power method, we would expect to see the opposite (short-dashed lines): favourites are overvalued, while underdogs are undervalued. We would then expect an estimated slope coefficient smaller than one and a positive intercept. This understanding is important, as it closely resembles what we observe in the data.

9.4 Testing the Informational Content of Betting Odds

9.4.1 Data

We use tennis match results and betting data as published on tennis-data.co.uk. The dataset covers results of the most important tennis tournaments in the period January 2001 until November 2018. Matches in this dataset are played by two individual players during a particular tournament, so matches between country teams are not included. The most important player-specific variables in the dataset are the name of the player, the ranking at the start of the current tournament, and number of points on the relevant world ranking (ATP for men, WTA for women). At the match level, the result is recorded (also in the case of a walk-over or retirement during the match) and the scores in each set. Also, betting odds, as offered by a maximum of eleven bookmakers for both players to win the match, are recorded. The odds are typically the most recent ones before the match starts, so they should incorporate all relevant information (injuries, form, recent results, etc.). Finally, at the tournament level we know the name and the location, the type of tournament (for example, Grand Slam, ATP500, or WTA International), and the surface of play.

In our empirical analysis we use betting data from the online bookmaker bet365.com, as these data are available for most matches. We retain observations with a maximum booksum of 1.15, which results in a dataset of 69117 tennis matches with available odds for both players.

9.4.2 Binned Data

The first method to test that probabilities implied by betting odds are unbiased estimators for outcome probabilities is based on binning the data into groups and comparing the average probability in the groups with the outcome frequencies. We divide the data in G groups, based on a division of the probability of player A into consecutive categories. For each match we randomly assign which player is player A.

Let the average probability of player A to win in group g be denoted by $\bar{p}_g$ and the actual win frequency of player A in group g by f_g. If probabilities are unbiased, the average price should be approx-

9.4. TESTING THE INFORMATIONAL CONTENT

imately equal to the actual win frequency, except for some random variation. This can be tested by estimating the regression equation

$$f_g = \beta_0 + \beta_1 \bar{p}_g + \varepsilon_g, \qquad (9.2)$$

with ε_g the random error term. Since the number of observations varies between groups, and since we estimate a fraction, the error term is heteroscedastic: $\text{var}(\varepsilon_g) = \sigma^2 \frac{\bar{p}_g(1-\bar{p}_g)}{n_g}$ with n_g the number of observations in group g. As a consequence, the parameters of equation (9.2) are estimated by weighted least squares. The hypothesis that probabilities are unbiased estimators of actual win probabilities is now tested by the joint hypothesis $\beta_0 = 0$ and $\beta_1 = 1$. After estimating equation (9.2), we do so by using a Wald-test.

Estimation Results Our testing procedure allows us to use either scaled probabilities or Shin probabilities as $\bar{p}_g$ in equation (9.2). We distribute the data in twenty groups based on the probability of player A, with each group of the same width. The result of this binning procedure is depicted in Figure 9.4. The left panel has scaled probabilities horizontally and winning frequencies vertically, the right panel has Shin probabilities horizontally and winning frequencies vertically. The size of the points reflects the number of observations in that group. For ease of interpretation, the relation between probabilities and win frequencies under the null hypothesis is drawn as a grey line.

It seems both probabilities provide a good fit to the null hypothesis, but on closer inspection it seems that for low values of scaled probabilities the actual win frequency is lower than the scaled probability, and for high values the actual win frequency is higher than the scaled probability. In other words, underdogs win less often than their scaled probability suggests, and favourites win more frequently. The scaled probabilities as a measure of win probability suffer from the favourite-longshot bias, as also documented in Forrest and McHale (2007). This is confirmed in the first two columns of Table 9.1, where we give the estimation results of equation (9.2). The Wald-statistic to test the hypothesis $\beta_0 = 0$ and $\beta_1 = 1$ is 74.108. Under the null hypothesis, this statistic follows a χ^2 distribution with 2 degrees of freedom, so the null hypothesis is rejected at any reasonable significance level. We conclude that scaled probabilities are not unbiased estimators of outcome probabilities in tennis.

The right panel of Figure 9.4 shows the relation between winning frequency and Shin probabilities. As discussed in the previous section, Shin probabilities correct for a potential favourite-longshot bias, and this is visible in that right panel: Shin probabilities are unbiased estimates of the actual win frequencies, also for small and large values of the price. The favourite-longshot bias seems to have disappeared. This is confirmed formally in the middle two columns of Table 9.1.

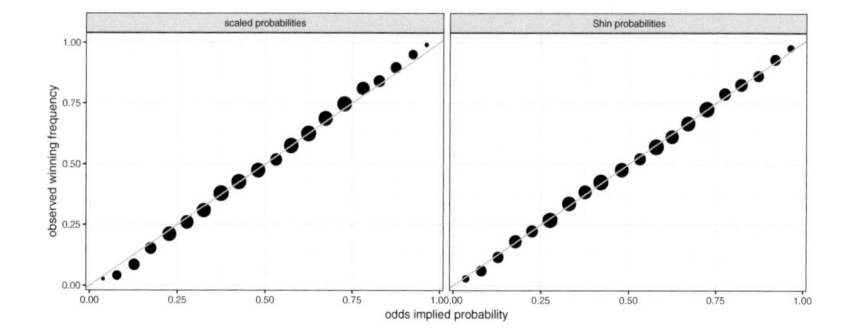

Figure 9.4: Observed winning frequencies against binned scaled probabilities (left) and Shin probabilities (right). The size of each observation is proportional to the number of observations in that bin.

Table 9.1: Point estimates and standard deviations (st.dev) of model (9.2), binned data. The p-value for the Wald statistic corresponding to the joint hypothesis $\beta_0 = 0$ and $\beta_1 = 1$ is given in the column with the standard deviations.

	scaled prob.		Shin prob.		power prob.	
	estimate	st.dev.	estimate	st.dev.	estimate	st.dev.
β_0	-0.037	0.003	-0.008	0.003	0.010	0.003
β_1	1.072	0.006	1.015	0.005	0.979	0.005
R^2	1.000		1.000		1.000	
W	74.108	0.000	3.584	0.167	14.512	0.001

9.4. TESTING THE INFORMATIONAL CONTENT

The p-value corresponding to the Wald-statistic (3.584) is 0.167. Shin probabilities are a better predictor of actual outcomes in tennis than scaled probabilities.

Results for power probabilities are given in the right two columns of Table 9.1 only. As expected from the discussion in the previous section and our estimation results so far, power probabilities show a reverse favourite-longshot bias: longshots win more frequently than expected based on the power probability. This is confirmed in the right two columns of Table 9.1: the estimated intercept is positive, and the slope is less than 1. Based on the Wald-statistic, we reject the null hypothesis: power probabilities are not unbiased estimators of actual outcome frequencies.

9.4.3 Individual Match Data

Our second testing strategy is based on individual matches. Suppose we consider a match between players A and B, and the odds implied probability of player A to win is p_A. The labels A and B are randomly assigned to either player. If this probability is an unbiased estimator, we should have

$$\Pr(A > B) = p_A,$$

with the left-hand side interpreted as the probability that A wins against B. We rewrite this probability as

$$\Pr(A > B) = p_A = \frac{1}{\frac{1}{p_A}} = \frac{1}{1 + \frac{1}{p_A} - 1} = \frac{1}{1 + \exp\left(\log\left(\frac{1}{p_A} - 1\right)\right)}. \tag{9.3}$$

To examine the empirical validity of this relation, we estimate

$$\Pr(A > B) = \frac{1}{1 + \exp\left(-\beta_0 - \beta_1 \log\left(\frac{1}{p_A} - 1\right)\right)} \tag{9.4}$$

and test whether $\beta_0 = 0$ and $\beta_1 = -1$ in which case equation (9.4) reduces to (9.3). This model is a simple logit model that is easily estimated using individual match data. Again, the central hypothesis $\beta_0 = 0$ and $\beta_1 = -1$ will be tested using a Wald-test. An attractive feature of this testing strategy compared to the one of the previous subsection is that it is straightforward to incorporate additional covariates. Variation in covariates between different matches would have averaged out in the binning approach of the previous subsection. If betting prices are fully informative, any additional covariate should have no effect on the outcome probability. Adding covariates, though, must be done carefully.

Suppose we have two players, labelled player A and player B. Again, note that the labels are arbitrary: one can think of tossing a coin before the match that determines which player is player A and which player is player B.

Then, the probability that player A wins against player B is, according to the logit model,

$$\Pr(A > B) = \frac{1}{1 + \exp(-f(x_A, x_B, z_{AB}))}. \tag{9.5}$$

In this expression, $f(x_A, x_B, z_{AB})$ is known as the index function, x_A are covariates specific to player A, x_B are covariates specific to player B, and z_{AB} are covariates common to both players A and B, but these vary between matches. Examples of player specific covariates are position on the world ranking and age. Examples of common covariates are type of tournament and surface of the match.

The fact that we randomly assign the label A to one of the players imposes the restriction on the index function that

$$-f(x_A, x_B, z_{AB}) = f(x_B, x_A, z_{AB}) \tag{9.6}$$

for all values of x_A, x_B and z_{AB}. To see this, consider two players: Federer and Nadal. The coin toss indicates that Federer is player A. We have according to (9.5)

$$\Pr(\text{Federer} > \text{Nadal}) = \frac{1}{1 + \exp(-f(x_{\text{Federer}}, x_{\text{Nadal}}, z_{\text{common}}))}. \tag{9.7}$$

What if the coin toss had been the other way around? In that case, player A would be Nadal, and player B would be Federer. Again using (9.5), we have

$$\Pr(\text{Nadal} > \text{Federer}) = \frac{1}{1 + \exp(-f(x_{\text{Nadal}}, x_{\text{Federer}}, z_{\text{common}}))}.$$

Since the probabilities associated with different outcomes need to add up to one, this also implies

$$\Pr(\text{Federer} > \text{Nadal}) = 1 - \frac{1}{1 + \exp(-f(x_{\text{Nadal}}, x_{\text{Federer}}, z_{\text{common}}))}$$
$$= \frac{1}{1 + \exp(f(x_{\text{Nadal}}, x_{\text{Federer}}, z_{\text{common}}))}. \tag{9.8}$$

Obviously, (9.7) and (9.8) should be equal, so we have the restriction

$$-f(x_{\text{Federer}}, x_{\text{Nadal}}, z_{\text{common}}) = f(x_{\text{Nadal}}, x_{\text{Federer}}, z_{\text{common}})$$

9.4. TESTING THE INFORMATIONAL CONTENT

for all values of x_{Federer}, x_{Nadal}, and z_{common}. This indeed corresponds to (9.6).

If we use the common linear index function:

$$f(x_A, x_B, z_{AB}) = \beta' x_A + \gamma' x_B + \delta' z_{AB},$$

by (9.6) we should have that

$$-\beta' x_A - \gamma' x_B - \delta' z_{AB} = \beta' x_B + \gamma' x_A + \delta' z_{AB},$$

or

$$\beta'(x_A + x_B) + \gamma'(x_A + x_B) + 2\delta' z_{AB} = (\beta + \gamma)'(x_A + x_B) + 2\delta' z_{AB} = 0.$$

If this is to hold for all values of the vectors x_A, x_B, and z_{AB}, we obtain the restrictions $\gamma = -\beta$ and $\delta = 0$, so that the index function takes the form

$$f(x_A, x_B, z_{AB}) = \beta'(x_A - x_B). \tag{9.9}$$

Only differences in person specific covariates and no common covariates can enter the index function. Intuitively, this makes sense: the fact that player 1 is better ranked than player 2 should increase his win probability, not just the numerical value of the ranking by itself. Furthermore, the fact that a match is played on hardcourt cannot increase *both* winning probabilities, as they must add up to 1.

Consistency requirement (9.9) is a simplified version of the ones discussed in Schauberger et al. (2018) and Schauberger and Tutz (2019). Their results reduce to equation (9.9) if the assumption of fixed slope coefficients β, γ, and δ is imposed. They allow for a more general specification, at the expense of having to use a more complex estimation method.

From the discussion above, it follows that common effects cannot enter the index function as main effects. The fact that a match is played on hard court cannot increase *both* winning probabilities. Still, it is possible to examine heterogeneity of the relation between winning probabilities and, say, surface type. The first approach is to estimate model (9.4) for each surface type, and then hypothesis $\beta_0 = 0$ and $\beta_1 = -1$ is tested simultaneously for these specifications. Below, we will follow this approach. In a second approach, performance of players on each surface can be measured (for example, their career winning percentage on clay, grass, and hard court), and differences in these measures between both players can be added as a covariate.

Estimation Results The estimation results of model (9.4) are given in Table 9.2, again for scaled, Shin, and power probabilities. The hypothesis $\beta_0 = 0$ and $\beta_1 = -1$ is soundly rejected for scaled prices. In other words, the relation $\Pr(A > B) = p_A$ does not hold for scaled prices. To assess the implication of the slope coefficient being significantly smaller than -1, we refer to the left panel of Figure 9.5. There,

Table 9.2: Point estimates and standard deviations (st.dev) of model (9.4), match level data. The p-value for the Wald statistic corresponding to the joint hypothesis $\beta_0 = 0$ and $\beta_1 = 1$ is given in the column with the standard deviations.

	scaled prob.		Shin prob.		power prob.	
	estimate	st.dev.	estimate	st.dev.	estimate	st.dev.
β_0	−0.003	0.009	−0.003	0.009	−0.003	0.009
β_1	−1.113	0.010	−1.015	0.009	−0.951	0.009
W	128.912	0.000	2.876	0.237	31.368	0.000

the estimated relation between the actual win probability (vertical) and the scaled price (horizontal) is drawn together with that relation under the null hypothesis (light grey line). For small values of the scaled price, the actual win probability is smaller than the price, and the reverse holds for large values of the scaled price. Again, the favourite-longshot bias implied by scaled prices is confirmed, here using individual match-level data.

As in the case of binned data, Shin prices are unbiased estimators of outcomes. This is shown in the right panel of Figure 9.5 and the middle two columns in Table 9.2. The p-value corresponding to the Wald statistic is 0.237 ($W = 2.876$, 2 degrees of freedom), so the null hypothesis is not rejected. Shin prices are unbiased estimators of tennis outcomes.

The results for the power probabilities are as before. In this specification, the slope coefficient being larger than −1 indicates a reverse favourite-longshot bias, and the point estimate for β_1 is significantly larger than −1 indeed. Consequently, the Wald-statistic indicates that the null hypothesis has to be rejected.

Now that we have shown that Shin probabilities are unbiased estimators of outcome probabilities, at least in the full dataset, we can look somewhat deeper by testing whether this holds for subgroups as well. We extend our test $\beta_0 = 0$ and $\beta_1 = -1$ by interacting the intercept and the $\log(\frac{1}{p_A} - 1)$ variable in equation (9.4) with dummy variables defining each subgroup. Our test then becomes a Wald-test of the hypothesis that the intercept is 0, the slope coefficient of $\log(\frac{1}{p_A} - 1)$ is −1, and the coefficients of all interactions are 0. The results are given in Table 9.3.

First, we test whether Shin probabilities are unbiased in both men and women matches. The p-value corresponding to this test is 0.338, so there is no statistical evidence that the prices either for men or

9.4. TESTING THE INFORMATIONAL CONTENT

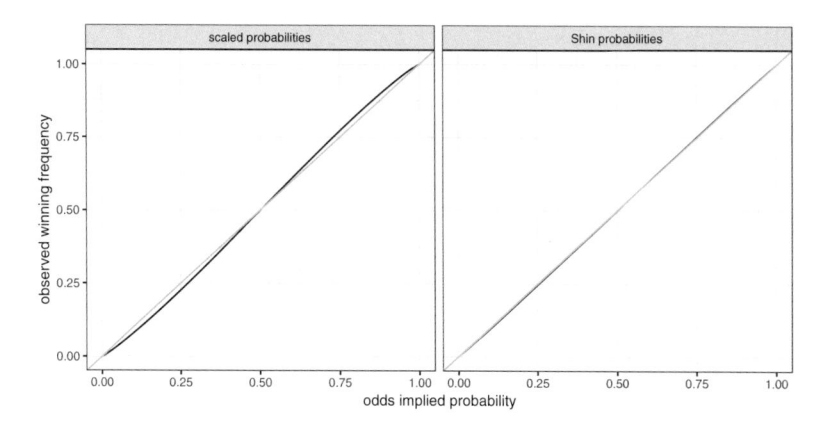

Figure 9.5: Win probability as a function of scaled probabilities (left) and Shin probabilities (right).

women matches are biased. Then we test a similar extension for Grand Slam vs. non-Grand Slam matches. The p-value is 0.001 in this case, and closer inspection shows that Shin prices are not unbiased for matches of Grand Slam tournaments ($p = 0.0001$) but they are unbiased for outcomes in non-Grand Slam matches ($p = 0.892$). The parameter estimates for Grand Slam matches ($\hat{\beta}_0 = -0.025$ and $\hat{\beta}_1 = -1.077$) are such that for this case the favourite-longshot bias reappears. This finding confirms the results in Lahvička (2014) and Abinzano et al. (2016). Forrest and McHale (2007) argue that this bias is due to the existence of bettors who are risk averse, but skewness loving. Considering the high profile of Grand Slam tournaments, there may be such a group of bettors who like to bet on underdogs hoping on a large return. Unfortunately, our data do not allow us to explore this further.

Next, we test whether type of surface is related to the unbiasedness of Shin probabilities. We distinguish three different types of surface: clay, grass, and hard court. Again, the p-value indicates that Shin probabilities are unbiased, irrespective of the surface of the match. Finally, it may be possible that Shin probabilities do not reflect a particular player being on an unexpected winning streak in a tournament. An unexpected winning streak can be measured by the return on bets in the previous rounds. If a player has won several matches unexpectedly, the pay-out on these matches would be high. The difference in realised returns on previous matches in each particular tournament is calculated for each player, and, following the identification discussion above, the difference between the returns of

Table 9.3: p-values of Wald tests that model (9.4) varies between levels of common covariates, match level data.

	p-value
sex	0.338
Grand Slam tournament	0.001
type of surface	0.505
return	0.314

both players is entered as an additional covariate in our model. The p-value of the hypothesis that that coefficient is 0 is 0.314, so there is no evidence that unexpected past performance in a given tournament is not priced into the Shin probabilities.

9.5 Practical Implications

To test the practical implications of the different probability estimates, we consider some simple betting strategies. The expected profit for a one-dollar bet on player A, when the odds are O_A and true winning probability is p_A, is equal to $p_A O_A - 1$. Likewise, for player B we have the expected profit $p_B O_B - 1$.

If we estimate the expected profit based on the scaled probabilities, by construction it is equal to $(O_A^{-1} + O_B^{-1})^{-1} - 1$ for both players. This means that using these probabilities, we cannot devise a betting strategy solely based on expected profit. However, if we substitute the Shin probabilities or the probabilities resulting from the power method, we generally obtain a preferred player on which we can then place a bet.

In general, we find that the overround is sufficiently large for the expected profit to be negative using either of the probabilities. We therefore consider a bettor who is willing to place bets when the estimated profit is above a certain threshold L, where L is negative.

We compare the following three strategies. In Strategy 1 we always bet on Player A. In Strategy 2 we always bet on the player with the highest expected profit, regardless of the value of the profit. Strategy 3 places a bet on the player with the highest expected profit, but only if this profit exceeds the (negative) threshold value L. Note that Strategy 2 is the same as Strategy 3 with $L = -\infty$. The data used is the same as before, with a total of 69117 potential bets.

9.5. PRACTICAL IMPLICATIONS

Strategy 1 is the benchmark strategy, where we always bet on Player A. The first row of Table 9.4 shows that based on the 69117 matches in the data, this leads to an average return of −9.5% per bet, and a total loss of 6550 dollars if we were to place one dollar bets.

Strategy 2 serves as a test whether we can use the Shin and power probabilities to increase the expected profit. Indeed, in Table 9.4, we see that Strategy 2 leads to a loss of 5.2% for both the Shin probabilities and the probabilities resulting from the power method. This is a notable difference with the 9.5% loss from Strategy 1. We further note that the choice for the winning player coincides for the Shin and power method probabilities. The lower amount of bets placed when using the power method results from the fact that the root solving algorithm underlying the power method does not always yield a solution. In this case, we do not place any bet.

Table 9.4: Comparison of betting strategy outcomes.

Strategy	L	Number of bets		% loss per bet		Total loss	
1	–	69117		-9.5		-6550	
		Shin	Power	Shin	Power	Shin	Power
2	$-\infty$	69117	69031	-5.2	-5.2	-3582	-3583
3	-0.050	37128	49616	-3.6	-4.2	-1332	-2087
	-0.025	2318	13801	-1.4	-2.3	-32	-324
	-0.010	15	2229	0.6	-1.9	0	-42

Note: Strategy 1: always bet on player A (benchmark). Strategy 2: bet on the player with the highest expected profit (or lowest expected loss) calculated using either Shin or power method probabilities. Strategy 3: bet on the player with the lowest expected loss only if this loss is smaller than L. Reported are the number of bets placed, the percentage loss per bet, and the total loss in dollars (assuming one-dollar bets).

The results for Strategy 3 are presented in the last three rows of Table 9.4. The second and third column show the number of bets that are taken for different values of L, the allowed loss per one dollar bet. For $L = -0.05$, we would still place a substantial amount of bets, but this rapidly decreases when we lower the allowed loss to -0.025. For $L = -0.01$, the use of Shin probabilities only suggests taking 15 bets. Conditional on taking a bet, the table shows the average loss per placed bet in columns four and five. We see that the Shin probabilities offer a lower loss per placed bet for the three choices of L. In the final two columns, we multiply the loss per bet (assuming a one dollar bet) with the number of bets taken. Here we see that the differences between the Shin probabilities and the power method probabilities can have quite sizeable economic consequences due to the large number of available bets.

9.6 Conclusion

This chapter concerned the information efficiency of tennis betting markets. We show that winning probabilities inferred through the model of Shin (1993) from a large data set on professional tennis players closely approximate the true winning probabilities. We compare this method to two other methods to deduce probabilities. The scaling method leads to estimated winning probabilities that are too low for heavy favourites and too high for underdogs. This is the well-known favourite-longshot bias. The power method is found to lead to reverse longshot bias. This can be explained under the assumption that the Shin (1993) model is indeed close to the underlying data generating process. The results are obtained using a binning technique, where matches with similar implied winning probabilities are grouped. We also consider a logit model to model individual matches and analyse the effects of several covariates. In agreement with the literature, the longshot bias is stronger in high profile (Grand Slam) events. This is the only case where we find evidence that the Shin probabilities are not unbiased estimators of the true winning probabilities.

Acknowledgement

Doreen Posthuma provided excellent research assistance.

Bibliography

Abinzano, I., Muga, L., and Santamaria, R. (2016). Game, set and match: the favourite-long shot bias in tennis betting exchanges. *Applied Economics Letters*, 23(8):605–608.

Boulier, B., Stekler, H., and Amundson, S. (2006). Testing the efficiency of the National Football League betting market. *Applied Economics*, 38(3):279–284.

Clarke, S., Kovalchik, S., and Ingram, M. (2017). Adjusting bookmaker's odds to allow for overround. *American Journal of Sports Science*, 5(6):45–49.

Del Corral, J. and Prieto-Rodríguez, J. (2010). Are differences in ranks good predictors for grand slam tennis matches? *International Journal of Forecasting*, 26(3):551–563.

Forrest, D., Goddard, J., and Simmons, R. (2005). Odds-setters as forecasters: The case of English football. *International Journal of Forecasting*, 21(3):551–564.

Forrest, D. and McHale, I. (2007). Anyone for tennis (betting)? *The European Journal of Finance*, 13(8):751–768.

Fort, R. (2011). *Sport Economics (Third Edition)*. Prentice-Hall, Upper Saddle River.

Goddard, J. and Asimakopoulos, I. (2004). Forecasting football results and the efficiency of fixed-odds betting. *Journal of Forecasting*, 23(1):51–66.

Golec, J. and Tamarkin, M. (1995). Do bettors prefer long shots because they are risk-lovers, or are they just overconfident? *Journal of Risk and Uncertainty*, 11(1):51–64.

Golec, J. and Tamarkin, M. (1998). Bettors love skewness, not risk, at the horse track. *Journal of Political Economy*, 106(1):205–225.

Klaassen, F. and Magnus, J. (2003). Forecasting the winner of a tennis match. *European Journal of Operational Research*, 148(2):257–267.

Koning, R. (2011). Home advantage in professional tennis. *Journal of Sports Sciences*, 29(1):19–27.

Lahvička, J. (2014). What causes the favourite-longshot bias? Further evidence from tennis. *Applied Economics Letters*, 21(2):90–92.

McHale, I. and Morton, A. (2011). A Bradley-Terry type model for forecasting tennis match results. *International Journal of Forecasting*, 27(2):619–630.

Pankoff, L. (1968). Market efficiency and football betting. *The Journal of Business*, 41(2):203–214.

Quandt, R. (1986). Betting and equilibrium. *The Quarterly Journal of Economics*, 101(1):201–207.

Sauer, R. (1998). The economics of wagering markets. *Journal of Economic Literature*, 36(4):2021–2064.

Sauer, R. (2005). The state of research on markets for sports betting and suggested future directions. *Journal of Economics and Finance*, 29(3):416–426.

Schauberger, G., Groll, A., and Tutz, G. (2018). Analysis of the importance of on-field covariates in the German Bundesliga. *Journal of Applied Statistics*, 45(9):1561–1578.

Schauberger, G. and Tutz, G. (2019). BTLLasso: A common framework and software package for the inclusion and selection of covariates in Bradley-Terry models. *Journal of Statistical Software*, 88(9):1–29.

Schnytzer, A., Lamers, M., and Makropoulou, V. (2010). The impact of insider trading on forecasting in a bookmakers' horse betting market. *International Journal of Forecasting*, 26(3):537–542.

Shin, H. (1993). Measuring the incidence of insider trading in a market for state-contingent claims. *Economic Journal*, 103(420):1141–1153.

Smith, M., Paton, D., and Williams, L. (2009). Do bookmakers possess superior skills to bettors in predicting outcomes? *Journal of Economic Behavior & Organization*, 71(2):539–549.

Stekler, H., Sendor, D., and Verlander, R. (2010). Issues in sports forecasting. *International Journal of Forecasting*, 26(3):606–621.

Štrumbelj, E. (2014). On determining probability forecasts from betting odds. *International Journal of Forecasting*, 30(4):934–943.

Sunde, U. (2009). Heterogeneity and performance in tournaments: a test for incentive effects using professional tennis data. *Applied Economics*, 41(25):3199–3208.

Vlastakis, N., Dotsis, G., and Markellos, R. (2009). How efficient is the European football betting market? Evidence from arbitrage and trading strategies. *Journal of Forecasting*, 28(5):426–444.

Vovk, V. and Zhdanov, F. (2009). Prediction with expert advice for the Brier game. *Journal of Machine Learning Research*, 10:2445–2471.

Williams, L. (1999). Information efficiency in betting markets: A survey. *Bulletin of Economic Research*, 51(1):1–39.

Williams, L. (2005). *Information Efficiency in Financial and Betting Markets*. Cambridge University Press.

Williams, L. and Paton, D. (1997). Why is there a favourite-longshot bias in British racetrack betting markets? *The Economic Journal*, 107(440):150–158.

Zuber, R., Gandar, J., and Bowers, B. (1985). Beating the spread: Testing the efficiency of the gambling market for National Football League games. *Journal of Political Economy*, 93(4):800–806.

Chapter 10
Fairness Trade-offs in Sports Timetabling

DRIES GOOSSENS
GHENT UNIVERSITY

XIAJIE YI
GHENT UNIVERSITY

DAVID VAN BULCK
GHENT UNIVERSITY

Abstract

Any sports tournament needs a timetable specifying which teams will meet at what time and where. However, not all sport timetables are equally fair for the contestants. In this chapter we discuss three fairness issues, namely consecutive home games, the carry-over effect (which relates to the opponent's previous game), and the number of rest days each team has between consecutive games. Since we typically cannot obtain a timetable that scores well on all these issues, we study how to make a good trade-off. Furthermore, we look at the trade-off between a timetable that is as fair as possible for the league overall versus a timetable that equitably splits its unfair aspects over the teams. We verify how several official timetables from major European football competitions score with respect to fairness criteria. Finally, we generate timetables for an amateur indoor football competition that reconcile overall fairness with an equitable distribution of unfairness over the teams.

10.1 Introduction

Every sports competition needs a timetable stating when and where each match of the tournament will be played. In professional sports the timetable has high economical stakes because it has an impact on commercial interests and revenues of clubs, broadcasters, sponsors, etc. In amateur sports, practical concerns like venue and team availability are more prominent. However, in all sports a fair timetable is paramount in the sense that it should not a priori give an advantage or a disadvantage to a particular contestant. Dealing with fairness, given the many stakeholders and the diversity of their (often conflicting) requirements, makes sports timetabling a very challenging optimisation problem (Van Bulck et al., 2020). We explain a number of essential concepts in sports timetabling in Section 10.2.1.

In this chapter we focus on so-called *round robin tournaments*. These are tournaments where each team meets each other team a fixed number of times, as opposed to, e.g., knock-out tournaments (used in, for example, Grand Slam tennis tournaments), where contestants do not face each opponent. The double round robin tournament, where each team faces each other team twice (typically once in its home venue and once in the venue of the opponent), is a particularly popular format. It is omnipresent, e.g., in football (Goossens and Spieksma, 2012b). While this tournament design alone already adds a substantial level of fairness to any timetable designed for it, there are many other fairness issues to deal with. In Section 10.2.2 we discuss breaks, the carry-over effect, and rest times as our main fairness criteria.

Unfortunately, there is no such thing as a free lunch (Friedman, 1975). Indeed, in sports timetabling, even though we prefer a timetable that is fair in every possible way, we are facing a trade-off between various fairness criteria. How much of one fairness criterion do we have to give up in order to make the timetable better with respect to some other fairness issue? Moreover, even if there was only one fairness criterion to deal with, the question remains how to fairly distribute the undesirable properties of the timetable over the teams. We explore some basic concepts on these fairness trade-offs in Section 10.2.3.

For academics we describe methods to minimise the number of breaks and the carry-over effect in Section 10.3. We develop a so-called first-carry-over, then-break approach in order to obtain a set of timetables that provide insight in this trade-off. We study how we can optimise the rest times of a timetable without overlooking the distribution of undesirable rest times over the teams. To this extent we develop a mathematical formulation as well as a bi-criteria evolutionary algorithm.

For practitioners we apply our ideas to two real-life settings in

Section 10.4. We study the official timetables of ten main professional European football leagues in order to understand how the breaks versus carry-over trade-off is made in practice. We create a set of timetables for an amateur indoor football league which illustrates the price of fairness: by how much do the overall rest times decrease if we wish to balance short rest times over the teams?

10.2 Preliminaries

10.2.1 Sports Timetabling

A game (or match) between team i and j, denoted $i - j$, means that team i plays at home, i.e., it uses its own venue (stadium) for that game, against away team j. A round is a set of games, often played on the same weekend, in which every team plays one game at most. A timetable essentially assigns a round to each match.

A timetable is said to be *time-constrained* if it uses the minimum number of rounds required to schedule all the games. In this chapter we assume that time-constrained competitions have an even number of teams. Consequently, each team plays exactly one game in each round in a time-constrained timetable. In contrast, *time-relaxed* timetables utilise (many) more rounds than there are games per team. If a team does not play in a round, it is said to have a *bye* in that round. Table 10.1 gives an example of a time-constrained timetable for eight teams (named **A** to **H**); a time-relaxed timetable for the same tournament is given in Table 10.2.

Time-constrained timetables are common practice in, e.g., professional football leagues in Europe (Goossens and Spieksma, 2012b). However, there are a few reasons why organisers may opt for a time-relaxed timetable. The main reason is its flexibility to consider venue or team availability constraints. Besides, it may simply be unattractive or even impractical to play multiple games simultaneously. The most extreme case example is an asynchronous timetable where at most one match takes place in each round (see Suksompong (2016)). Asynchronous tournaments occur when there is only one venue, as is the case in the top-tier national football league of Gibraltar, or when fans desire to be able to watch all games live (e.g., the 2012 Premier League Snooker in England).

When the second half of a double round robin timetable is identical to the first half, except that the home advantage is flipped, we say that the timetable is *mirrored*.

10.2.2 Fairness Criteria

In this section we discuss three fairness criteria: breaks, the carry-

Table 10.1: A time-constrained single round robin timetable with eight teams.

r1	r2	r3	r4	r5	r6	r7
A-H	C-A	A-E	G-A	A-B	D-A	A-F
B-G	H-B	D-B	B-F	C-G	B-C	E-B
F-C	G-D	C-H	E-C	F-D	G-E	C-D
D-E	E-F	F-G	H-D	E-H	H-F	G-H

Table 10.2: A time-relaxed single round robin timetable with eight teams.

r1	r2	r3	r4	r5	r6	r7	r8	r9	r10
A-H	B-G	C-A	A-E	G-A	E-C	A-B	F-D	D-A	A-F
	F-C	H-B	D-B	B-F	H-D	C-G	E-H	B-C	E-B
	D-E	G-D	C-H					G-E	C-D
		E-F	F-G					H-F	G-H

over effect, and rest times. In our opinion these criteria are quite universal in the sense that in most sports, league organisers, teams, and fans would like to see them respected. However, we do not claim that there are no other universal fairness criteria, and there certainly are relevant sport-specific fairness criteria as well.

Breaks

Determining the venue of games is crucial in terms of the fairness of a timetable. The sequence of home matches ("H") and away matches ("A") played by a single team is called its *home-away pattern* (HAP). Given such a HAP, the occurrence of two consecutive home matches, or two consecutive away matches is called a *break*. Teams can have consecutive breaks, causing them to play three or more home (away) games in a row.

In some cases away breaks may be beneficial. For instance, to reduce travel costs, a team may prefer to have two (or more) consecutive away games if its stadium is located far from the opponents' venues, and the venues of these opponents are close to each other (provided of course that the teams do not return home after each game). However, in most competitions, breaks–and successive breaks in particular–are avoided as much as possible. Indeed, Forrest and Simmons (2005) observe that scheduling consecutive home games has a negative impact on attendance. Moreover, given the home advantage, a morale boost (blow) after two consecutive home (away) games may have an impact on the outcome of the next game. Thus, breaks are considered unfair.

10.2. PRELIMINARIES

For the sake of illustration, Table 10.3a gives the HAPs corresponding to the timetable presented in Table 10.1. Note that the HAPs of all teams except for team **A** and **H** contain a break. The timetable depicted here contains 6 breaks in total.

The Carry-Over Effect

Any timetable implies an order in which each team meets its opponents. Playing against a strong or a weak opponent has impact on the performance of teams (Briskorn and Knust, 2010). For instance, a team is more likely to be exhausted or demoralised, or to suffer injuries or suspensions from playing against a very strong opponent, which in turn can have a negative impact on this team's performance in its next game. In this way the opponent of this team in the next game receives an indirect advantage from that strong team. Following this idea, we say that a team i gives a *carry-over effect* to a team j if some other team t's game against i is immediately followed by a game of t against j.

Clearly, carry-over effects cannot be avoided as–except in the first round–teams always have a previous opponent. It is, however, considered unfair if some team predominantly gives a carry-over effect to the same team. Indeed, Goossens and Spieksma (2012a) mention examples from football in Norway and Belgium, where the carry-over effect was held responsible for determining the league champion and the relegated team (they could, however, not find any statistical evidence in favour of this claim).

The extent to which carry-over effects are balanced is measured by the so-called *carry-over effects value* (Russell, 1980). The carry-over effects value is defined as $\sum_{i,j} c_{ij}^2$, where c_{ij} corresponds to the number of times that team i gives a carry-over effect to team j in a tournament, i.e., the number of times that team j plays against the opponent of team i in the previous round. Note that, according to Russell's definition, carry-over effects from the last round to the first round are also counted. An illustration of the carry-over effects (c_{ij}) of the timetable in Table 10.1 is given in Table 10.3b. In this case the carry-over effects value equals 196.

Rest Times

Since time-relaxed timetables contain (many) more rounds than games per team, the rest time between consecutive games of a team can vary substantially, and a team's timetable may therefore contain congested periods. From a fairness perspective, congested periods are problematic as they can lead to player injuries (Bengtsson et al., 2013). For example, when Manchester City had to play four games in 11 days during the Christmas and New Year period in the 2018-2019 Premier League season, their coach Pep Guardiola said "the Premier League

Table 10.3: Illustration of HAPs and c_{ij} values of the timetable in Table 10.1.

(a) HAPs

	r1	r2	r3	r4	r5	r6	r7
A	H	A	H	A	H	A	H
B	H	A	A	H	A	H	A
C	A	H	H	A	H	A	H
D	H	A	H	A	A	H	A
E	A	H	A	H	H	A	H
F	H	A	H	A	H	A	A
G	A	H	A	H	A	H	H
H	A	H	A	H	A	H	A

(b) c_{ij} values

	A	B	C	D	E	F	G	H
A	0	1	5	0	0	0	0	1
B	0	0	1	5	0	0	0	1
C	0	0	0	1	5	0	0	1
D	0	0	0	0	1	5	0	1
E	0	0	0	0	0	1	5	1
F	5	0	0	0	0	0	1	1
G	1	5	0	0	0	0	0	1
H	1	1	1	1	1	1	1	0

fixture list is a disaster for player well being". West Ham's head of medical services Gary Lewin added: "I don't think it is particularly fair–physically it is not a level playing field for all clubs, as some are playing every few days, some have a longer break between games and some have a week off before playing twice in three days"[1].

Suksompong (2016) proposes to measure timetable congestion by the so-called *guaranteed rest time* (GRT), i.e., the number of rounds without a game that any team will at least have between two consecutive games. However, the problem with the GRT is that it only considers the worst-case rest time. Therefore, in this chapter, we use the *aggregated rest time penalty* (ARTP) as a fairness measure. This measure penalises a timetable with a value p_r each time a team has only r rounds between two consecutive games. The idea is, of course, that this penalty increases as there are fewer rounds in between; on the other hand, when sufficient rounds are in between such that the team is fully recovered, no penalty is incurred (Van Bulck et al., 2019). The ARTP of a timetable is then simply the sum of these penalties.

As an illustration, Table 10.4 gives the calculation of the rest time penalties for the timetable depicted in Table 10.2. If we assume $p_0 = 4$, $p_1 = 2$ and $p_2 = 1$, we have ARTP = 157. Although the GRT of any timetable with eight teams and 10 rounds is always 0, there are several

---

[1] CNN Sports, https://edition.cnn.com/2018/01/02/football/pep-guardiola-english-premier-league-fixture-scheduling/index.html

10.2. PRELIMINARIES

Table 10.4: Calculation of the rest time penalties for the timetable in Table 10.2.

	r1	r2	r3	r4	r5	r6	r7	r8	r9	r10
A			p_1	p_0	p_0		p_1		p_1	p_0
B			p_0	p_0	p_0		p_1		p_1	p_0
C			p_0	p_0		p_1	p_0		p_1	p_0
D			p_0	p_0		p_1		p_1	p_0	p_0
E			p_0	p_0		p_1		p_1	p_0	p_0
F			p_0	p_0	p_0			p_2	p_0	p_0
G			p_0	p_0	p_0		p_1		p_1	p_0
H			p_1	p_0		p_1		p_1	p_0	p_0

opportunities to optimise the ARTP.

10.2.3 Fairness Trade-Offs

Unfortunately, a single timetable that gives the best performance for each fairness criterion does not exist. Consequently, league organisers are facing a trade-off between fairness criteria. One way to organise this trade-off is through *Pareto-efficiency*. A timetable is called Pareto-efficient if it is impossible to improve one fairness criterion without deteriorating at least one other fairness criterion. This concept is named after the Italian economist and engineer Vilfredo Pareto (1848–1923), who applied it in his income distribution studies. Hence, it would be wise to pick a timetable from the set of Pareto-efficient timetables, known as the *Pareto-front*. The choice for a specific timetable on the Pareto-front depends on the relative importance the league organiser attaches to the fairness criteria.

Even if the league organiser manages to strike a balance between multiple fairness criteria, his struggle is not over. Indeed, the previously described approach only considers the overall presence of unfairness; it does not consider how this unfairness is distributed over the teams. Even a timetable with very few undesired properties may be conceived as unfair when only a small number of teams carry most of this burden. We assume that each undesired property of a timetable can be linked with a penalty, which, in turn, can be split over the teams that suffer from it. Thus, we need to reconcile *criterion efficiency*, i.e., minimising the total penalty of the timetable, with *distribution equity*, to make sure that the distribution of the penalties is well balanced over the teams.

We aim to generate equitably-efficient timetables. Practically, a timetable is equitably-efficient if it is Pareto-efficient, meaning that we cannot improve a team's timetable without deteriorating the timetable of at least one other team, and it complies with the Pigou-Dalton principle of fairness, meaning that we cannot shift a penalty from a worse-off team to a better-off team without increasing the total penalties (Ogryczak, 1997). Equitable-efficiency can be seen as a refinement of Pareto-efficiency: any equitably-efficient timetable is Pareto-efficient, but not all Pareto-efficient timetables are equitably-efficient.

10.3 For Academics

10.3.1 Trade-Offs Between Two Fairness Criteria

In time-constrained timetables for round robin tournaments with an even number of teams, each team will play on each round, and, consequently, their rest times will be the same. Therefore, in these competitions, we focus on minimising breaks and balancing the carry-over effect. We survey what is known in the literature on these fairness criteria, as well as their trade-off, before we develop our own approach.

Minimising the Number of Breaks

Many of the theoretical results and algorithms in sports timetabling are based on graph theory. For instance, De Werra (1980) uses the complete graph K_{2n} on $2n$ nodes for constructing single round robin tournaments, with the nodes corresponding with the teams and each edge with a game between the teams of the nodes it connects. A time-constrained timetable can then be seen as a one-factorisation of K_{2n}, i.e., a partitioning into edge-disjoint one-factors F_i with $i = 1, ..., 2n - 1$. A one-factor is a set of edges such that each node in the graph is incident to exactly one of these edges. Each one-factor corresponds to a round and represents n matches.

One particular one-factorisation results in so-called *canonical* timetables, which are highly popular in sport timetabling (Goossens and Spieksma, 2012b). This one-factorisation has its one-factors F_i for $i = 1, ..., 2n - 1$ defined as

$$F_i = \{(2n, i)\} \cup \{([i+k]^+, [i-k]^-) : k = 1, ..., n-1\},$$

where $[x]^+ = x$ if $x \leq 2n - 1$ and $[x]^+ = x - 2n + 1$ otherwise, while $[x]^- = x$ if $x \geq 1$ and $[x]^- = x + 2n - 1$ otherwise. Figure 10.1 illustrates the canonical one-factorisation for a single round robin tournament with 6 teams.

If the league organiser can determine which match is played in which round, the minimal number of breaks for a single-round robin tournament with $2n$ teams is $2n - 2$, with $2n - 2$ teams having 1 break and 2 teams without breaks (De Werra, 1981). Moreover, De Werra (1981) shows that this can always be achieved with a canonical

10.3. FOR ACADEMICS

timetable. For a double-round robin tournament, a timetable with $2n - 4$ breaks can easily be constructed from a single-round robin timetable with a minimal number of breaks; if we want a mirrored double-round robin timetable, the minimal number of breaks is $3n - 6$ (De Werra, 1981). However, if there is no need for a timetable that consists of consecutive single-round robin tournaments, we can limit the number of breaks to $n - 2$, even if all teams meet each other team more than twice (Goossens and Spieksma, 2012b).

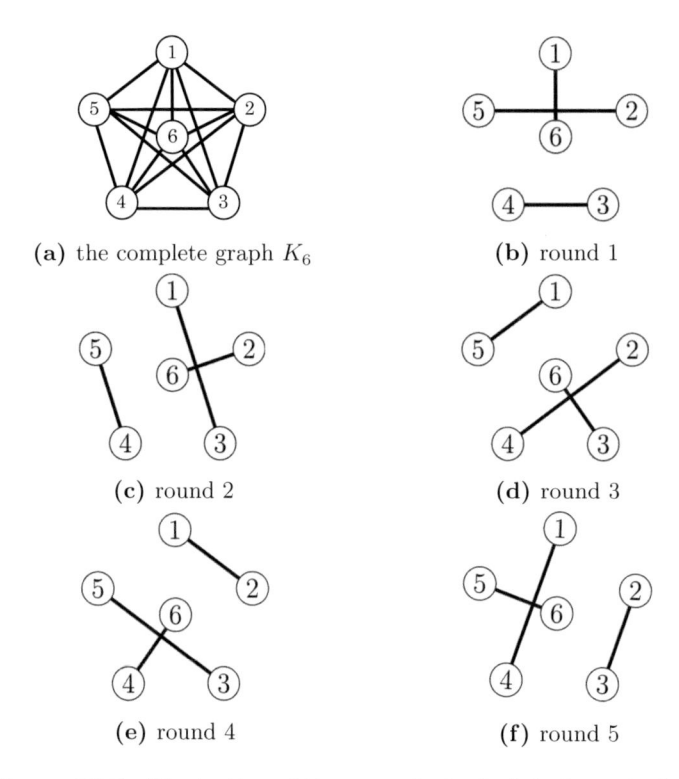

(a) the complete graph K_6

(b) round 1

(c) round 2

(d) round 3

(e) round 4

(f) round 5

Figure 10.1: Illustration of the canonical schedule for a single round robin tournament with 6 teams (based on Januario et al. (2016)).

If the opponents are fixed for each round, and the league organiser can only determine the home advantage, finding a timetable with a minimal number of breaks is known as the *break-minimisation problem*. This problem has been tackled using, e.g., constraint programming (Régin, 2001), integer programming (Trick, 2000) and semidefinite programming (Miyashiro and Matsui, 2006).

Table 10.5: Carry-over effects values found in the literature for various league sizes (proven optimal values are bolded).

$2n$	$2n(2n-1)/$ Best found	Russell	Anderson	Trick	Henz et al.	Miyashiro et al.	Guedes & Ribeiro	Kidd
4	12 / **12**	12	-	-	-	-	12	12
6	30 / **60**	60	-	60	60	-	60	60
8	56 / **56**	56	56	-	-	-	56	56
10	90 / **108**	138	108	122	128	108	108	108
12	132 / 160	196	176	-	188	176	160	176
14	182 / 234	260	234	-	-	254	254	234
16	240 / **240**	240	-	-	-	240	-	240
18	306 / 340	428	340	-	-	400	-	340
20	380 / **380**	520	380	-	-	488	-	380
22	462 / **462**	-	462	-	-	-	-	462
24	552 / 598	-	644	-	-	-	-	598

Balancing the Carry-Over Effect

Since its introduction by Russell (1980), there have been several attempts to find timetables with minimal carry-over effects value (or in other words, to balance the carry-over effects as evenly as possible over the teams). The lowest carry-over effects value that we may hope for in a single-round robin tournament with $2n$ teams is $2n(2n-1)$. This is the case when each team gives a carry-over effect to each other team (except itself) exactly once. A timetable that achieves this is called a *balanced* timetable.

Russell (1980) presents an algorithm that results in a balanced timetable when the number of teams is a power of 2. Anderson (1999) found balanced timetables for 20 and 22 teams and improved several of Russell's results. In fact, despite various approaches (Henz et al., 2004; Kidd, 2010; Miyashiro and Matsui, 2006; Trick, 2000), only Guedes and Ribeiro (2011) were able to further improve one of Anderson's results (i.e., for 12 teams). Trick (2000) proved that 60 is the optimal carry-over effects value for 6 teams; Lambrechts et al. (2017) later showed that all timetables for 6 teams have this carry-over effects value. Table 10.5 summarises the results.

Breaks Versus Carry-Over Effects

The best-known timetables with respect to the carry-over effects value do not specify the home advantage and make no claims about the number of breaks. On the other hand, the canonical timetables, which allow a minimal number of breaks, have been shown to be

10.3. FOR ACADEMICS

the worst timetables with respect to balancing the carry-over effects (Lambrechts et al., 2017). The timetable in Table 10.1 is canonical; its carry-over effects value amounts to 196 (see Table 10.3b).

The trade-off between the carry-over effect and breaks has been studied before in the literature. Çavdaroğlu and Atan (2019) start from the canonical timetable and apply a round swapping procedure to reduce its carry-over effects value. Günneç and Demir (2019) start from a timetable with one break per team at most, and then swap rounds using a tabu-search algorithm to obtain a small carry-over effects value. These contributions can be viewed as *first-break, then carry-over* approaches. Motivated by this idea, and taking into account that carry-over effects are only related to the opponents, here, we develop a reversed approach: minimising the carry-over effects value first and then optimising the number of breaks, or in other words *first-carry-over, then-break* (FCTB).

We first match the opponent pairs to rounds to minimise the carry-over effects value. Let T and R be the set of teams and rounds respectively, with $|T|$ even, and $|R| = |T| - 1$ (i.e., time-constrained single round robin). We say that x_{ijr} equals 1 if teams i and j $(i, j \in T, i \neq j)$ play against each other in round $r \in R$, and 0 otherwise. Next, we say that c_{ijr} equals 1 if team $i \in T$ gives a carry-over effect to team $j \in T$ in round $r \in R$, and 0 otherwise. As a result, the number of carry-over effects c_{ij} that team i gives to team j is the sum of c_{ijr} over $r \in R$. Thus, we obtain the following formulation (note that the carry-over effects from the last round to the first round are also considered):

$$\min \sum_{i \in T} \sum_{j \in T} c_{i,j}^2 \tag{10.1}$$

subject to

$$x_{i,i,r} = 0 \qquad \forall i \in T, r \in R \tag{10.2}$$

$$x_{i,j,r} = x_{j,i,r} \qquad \forall i,j \in T, r \in R \tag{10.3}$$

$$\sum_{r \in R} x_{i,j,r} = 1 \qquad \forall i,j \in T, i \neq j \tag{10.4}$$

$$\sum_{j \in T} x_{i,j,r} = 1 \qquad \forall i \in T, r \in R \tag{10.5}$$

$$x_{i,l,r} + x_{j,l,r+1} - 1 \leqslant c_{i,j,r} \qquad \forall i,j,l \in T, r \in R \setminus \{|R|\} \tag{10.6}$$

$$x_{i,l,|R|} + x_{j,l,1} - 1 \leqslant c_{i,j,|R|} \qquad \forall i,j,l \in T \tag{10.7}$$

$$\sum_{r \in R} c_{i,j,r} = c_{i,j} \qquad \forall i,j \in T, i \neq j \tag{10.8}$$

$$x_{i,j,r} \in \{0,1\} \qquad \forall i,j \in T, r \in R \tag{10.9}$$

$$c_{i,j,r} \in \{0,1\} \qquad \forall i,j \in T, i \neq j, r \in R \tag{10.10}$$

$$c_{i,j} \geqslant 0 \qquad \forall i,j \in T, i \neq j \quad (10.11)$$

Constraints (10.2) state that a team cannot play against itself, and constraints (10.3) enforce that if team i meets team j on some round that j then also meets i on that round. Constraints (10.4) regulate that each pair of teams meets once during the tournament and constraints (10.5) imply that each team plays exactly one game per round. Constraints (10.6) and (10.7) calculate the carry-over effects from team i to j over the tournament, including the effects passed from the last round to the first round. The relations between c_{ijk} and c_{ij} are shown in constraints (10.8).

After obtaining the assignment of opponent pairs to rounds with minimised carry-over effects from the formulation above (however, without the home advantage), the number of breaks is to be minimised. We use the following two decision variables: h_{ir} is 1 if team $i \in T$ plays home in round $r \in R$, and 0 otherwise, and b_{ir} equals 1 if team $i \in T$ has a break in round $r \in R$, and 0 otherwise. Note that x_{ijr} is known from the outcome of the before-mentioned formulation.

$$\min \sum_{i \in T} \sum_{r \in R} b_{i,r} \qquad (10.12)$$

subject to

$$2 - h_{i,r} - h_{j,r} \geqslant x_{i,j,r} \qquad \forall i,j \in T, r \in R \quad (10.13)$$

$$h_{i,r} + h_{j,r} \geqslant x_{i,j,r} \qquad \forall i,j \in T, r \in R \quad (10.14)$$

$$h_{i,r} + h_{i,r+1} - 1 \leqslant b_{i,r+1} \qquad \forall i \in T, r \in R \setminus \{|R|\} \quad (10.15)$$

$$1 - h_{i,r} - h_{i,r+1} \leqslant b_{i,r+1} \qquad \forall i \in T, r \in R \setminus \{|R|\} \quad (10.16)$$

$$\sum_{r \in R} h_{i,r} \leqslant |T|/2 \qquad \forall i \in T \quad (10.17)$$

$$\sum_{r \in R} h_{i,r} \geqslant |T|/2 - 1 \qquad \forall i \in T \quad (10.18)$$

$$h_{i,r}, b_{i,r} \in \{0,1\} \qquad \forall i \in T, r \in R \quad (10.19)$$

According to constraints (10.13), two teams who play against each other cannot both play home simultaneously nor can they both have an away game on that round (constraints (10.14)). Constraints (10.15) and (10.16) keep track of the home and away breaks, respectively. Additionally, the number of home and away games for each team should also be balanced, as enforced by constraints (10.17) and (10.18).

In our FCTB approach the idea is to first solve formulation (10.1)-(10.11), and then use its outcome as the input for formulation (10.12)-(10.19). The models are solved using IBM Ilog Cplex 12.71 on a MacOS 10.13.6 system with 8 GB RAM and an Intel Core i5 processor. For practical reasons we bound the computation time to 2 hours for

10.3. FOR ACADEMICS

Table 10.6: Trade-offs between breaks and carry-over effects value.

Number of teams		4	6	8	10	12	14	16	18	20	22	24
Canonical timetable	breaks	2	4	6	8	10	12	14	16	18	20	22
	carry-over	12	60	196	468	924	1612	2580	3876	5548	7644	10212
FCTB	breaks	2	4	8	12	18	24	34	40	-	-	-
	carry-over	12	60	56	138	206	292	400	508	-	-	-
FCTB-best-known	break	2	4	8	14	22	28	24	48	50	58	64
	carry-over	12	60	56	108	176[2]	234	240	340	380	462	598
Çavdaroğlu & Atan	breaks	-	-	-	12	18	26	32	42	-	-	-
	carry-over	-	-	-	136	192	254	330	406	-	-	-
Günneç & Demir[1]	breaks	-	-	-	12	16	24	34	40	-	-	-
	carry-over	-	-	-	144	212	302	396	556	-	-	-

[1] These results are from the setting with unbounded breaks.
[2] The best carry-over effects value for 12 teams is 160 (Guedes and Ribeiro, 2011); however, we were not able to retrieve the corresponding timetable from the literature.

solving formulation (10.1)-(10.11). For up to 8 teams, this suffices for finding an optimal solution. For more teams we check whether further improvements are possible by randomly switching the order of the rounds. As an alternative approach, we skip formulation (10.1)-(10.11) and instead use the best-known timetables with respect to the carry-over effects value published by Kidd (2010) as input for formulation (10.12)-(10.19). We refer to this approach as FCTB-best-known.

The results for the canonical timetable from the literature and from both our approaches for 4 to 24 teams are shown in Table 10.6. For 4 and 6 teams, FCTB obtains the same results as the canonical timetables, and both the carry-over effects value and the number of breaks are optimal. In comparison with the results from Çavdaroğlu and Atan (2019), the excellent carry-over effects values obtained by FCTB-best-known come at the cost of a slightly higher number of breaks (except for 16 teams). In most cases we have better solutions on both carry-over effects values and the number of breaks than the result from Günneç and Demir (2019). The timetables that are on the Pareto-front are indicated in bold. Note that the number of breaks per team is not bounded in our approach, but we enforce a balance in the number of home and away games played by every team during the tournament which is not considered in the literature.

10.3.2 Trade-Offs Between Efficiency and Equity

Typically, there are no timetables that make every team happy. Indeed, for each timetable $D \in \mathcal{D}$, team $1 \leqslant i \leqslant n$ has a non-negative aversion $f_i(D) : \mathcal{D} \to \mathbb{R}^+$. This aversion is based on one or more properties of the timetable (e.g., a break) that is undesirable for team i. We assume that the league organiser has perfect knowledge of the aversion of each team and is in control to select a timetable D from $\mathcal{D}$.

As it is in the league organiser's interest to please as many teams as possible, each function f_i needs to be minimised. A traditional approach is to use a *minisum* objective $\sum_{i=1}^{n} f_i(D)$, which minimises the total (or equivalently, the mean) aversion to the timetable. The main flaw of this approach is that only the aggregate of the undesirable property is considered (i.e., *criterion efficiency*) and not the distribution of the property over the teams (i.e., *distribution equity*).

In recent years the literature has come up with various inequity-averse optimisation techniques that incorporate distribution equity alongside criterion efficiency (see Karsu and Morton (2015) for an overview). Of particular interest is the concept of *equitable-efficiency* (Ogryczak, 1997). To explain this concept, we need some additional terminology. Aversion vector $\vec{x}_D = (f_1(D), f_2(D), \ldots, f_n(D))$ collects the aversion of all teams. We assume that $\vec{x}_D$ is sorted in non-increasing order meaning that $f_1(D) \geqslant f_2(D) \geqslant \cdots \geqslant f_n(D)$. As the league organiser strives for maximal efficiency, less aversion for one team is always better: $\vec{x}_D$ *rationally dominates* $\vec{x}_{D'}$ if the aversion for at least one team is smaller whereas no other team is worse off. We call $\vec{x}_D$ *Pareto-efficient* in $\mathcal{D}$ if, and only if, no timetable $D' \in \mathcal{D}$ exists such that $\vec{x}_{D'}$ rationally dominates $\vec{x}_D$. The image of all non-dominated solutions in the objective space is called the *Pareto-frontier*.

In order to fairly distribute the total aversion over the teams, the concept of equitable-efficiency is built around two fundamental axioms. The first axiom prescribes *anonymous identities*, i.e., the identities of the teams are not important to analyse the equity of the aversion vector. Although this seems reasonable, the practice of sports timetabling may be different in the sense that some teams can be considered *more equal* than others. The second axiom is known as the *Pigou-Dalton principle of transfers*: any transfer from a worse-off team to a better-off team, *ceteris paribus*, should always result in a more preferable aversion vector. For a formal definition of the axioms, we refer to Ogryczak (2000).

Vector $\vec{x}_D$ *equitably dominates* vector $\vec{x}_{D'}$ if we can produce $\vec{x}_D$ from $\vec{x}_{D'}$ after a finite sequence of index permutations and at least one aversion transfer from a worse-off team to a better-off team (also called a Robin Hood operation) or a decrease of the aversion of a particular team. We call $\vec{x}_D$ *equitably-efficient* if, and only if, no timetable $D' \in \mathcal{D}$ exists such that $\vec{x}_{D'}$ equitably dominates $\vec{x}_D$. As both axioms do not conflict with Pareto-efficiency, an equitably-efficient timetable is also Pareto-efficient but not the other way around (Karsu and Morton, 2015).

The remainder of this section is as follows. First, we describe a typical time-relaxed sports timetabling problem in which venue and

10.3. FOR ACADEMICS

player availability constraints need to be taken into account. Next, we propose an integer programming (IP) model to generate equitably-efficient timetables with regard to teams' rest periods. This IP model, however, requires a considerable amount of computational resources. Finally, we present a multi-objective evolutionary algorithm capable of generating a rich set of equitably-efficient solutions in a single run.

Availability Constraints in Time-Relaxed Timetabling

In the time-relaxed availability constrained double round-robin tournament (RAC-2RRT) problem, the input consists of a set of rounds R and a set of teams T where $n = |T|$. Each team $i \in T$ also provides a venue availability set $H_i \subseteq R$ containing all rounds during which i's venue is available and a player availability set A_i containing all rounds during which i's players are available. Since a team can only play (at home or away) when its players are available, we assume, without loss of generality, that $H_i \subseteq A_i$ for each $i \in T$. This makes that a team can play at home on all rounds in H_i, and that it can play away on all rounds in A_i. Finally, a parameter τ is given, defining the total number of rounds after which we assume a team is fully recovered from its previous game (e.g., 5 days).

RAC-2RRT consists of finding a feasible timetable that is an assignment of games to rounds such that:

(C1) each team plays exactly one home game against every other team,

(C2) the venue availability H_i with $i \in T$ is respected (i.e., no game i-j is planned on a round $r \notin H_i$),

(C3) the player availability A_i with $i \in T$ is respected (i.e., no game i-j or j-i is planned on a round $r \notin A_i$),

(C4) each team plays at most one game per round $r \in R$, and

(C5) each team plays at most two games per $\tau + 1$ rounds.

Van Bulck et al. (2019) propose IP formulation (10.20)-(10.24) to solve RAC-2RRT. In this model the variable $z_{i,j,r}$ is 1 if team $i \in T$ and team $j \in T \setminus \{i\}$ meet at the venue of i on round $r \in R$. Constraints (10.20) ensure that each team plays exactly one home game against every other team *(C1)*. Constraints (10.21) require a team to play at most one game per round *(C4)*, whereas constraints (10.22) enforce that a team plays at most two games per $\tau + 1$ rounds *(C5)*. Constraints (10.23) reduce the number of variables in the system by explicitly stating that two teams can only meet when the venue of the home team and the players of the away team are simultaneously available *(C2)*, *(C3)*; in practice these variables are not created. Finally, constraints (10.24) are the binary constraints on the z-variables.

$$\sum_{r \in H_i \cap A_j} z_{i,j,r} = 1 \qquad \forall i,j \in T : i \neq j \quad (10.20)$$

$$\sum_{j \in T \setminus \{i\}} (z_{i,j,r} + z_{j,i,r}) \leqslant 1 \qquad \forall i \in T, r \in A_i \quad (10.21)$$

$$\sum_{j \in T \setminus \{i\}} \sum_{k=r}^{r+\tau} (z_{i,j,k} + z_{j,i,k}) \leqslant 2 \qquad \forall i \in T, r \in A_i \quad (10.22)$$

$$z_{i,j,r} = 0 \qquad \forall i,j \in T : i \neq j, r \in R \setminus \{H_i \cap A_j\} \quad (10.23)$$

$$z_{i,j,r} \in \{0,1\} \qquad \forall i,j \in T : i \neq j, r \in H_i \cap A_j \quad (10.24)$$

The ARTP penalises the timetable with a positive value p_r each time a team has only $r < \tau$ rounds between consecutive games with $p_r \leqslant p_{r-1}$. We measure the aversion of team i by summing over the rest time penalties related to all games of team i and refer to this sum with ARTP_i. To minimise the ARTP, the minisum model uses an auxiliary variable $y_{i,r,t}$ which is 1 if team i plays a game on round r followed by its next game on round t, and 0 otherwise. Constraints (10.25) regulate the value of the $y_{i,r,t}$ variables by considering the number of rounds between two consecutive games of the same team. From *(C5)* it follows that the games are consecutive if team i plays on round r and t and $|t - r| \leqslant \tau$. We can strengthen the formulation by dropping the negative summation term of Equation (10.25). Constraints (10.26) model the aversion of team i by setting f_i equal to ARTP_i. Finally, constraints (10.27) state that the y-variables are binary.

Minisum model

$$\min \sum_{i \in T} f_i$$

subject to

$(10.20) - (10.24)$

$$\sum_{j \in T \setminus \{i\}} \left(z_{i,j,r} + z_{j,i,r} + z_{i,j,t} + z_{j,i,t} \right.$$
$$\left. - \sum_{k=r+1}^{t-1} (z_{i,j,k} + z_{j,i,k}) \right) - 1 \leqslant y_{i,r,t} \quad \forall i \in T, r, t \in A_i : r < t, t - r \leqslant \tau \quad (10.25)$$

$$f_i = \sum_{r \in A_i} \sum_{t=r+1}^{r+\tau} p_{(t-r-1)} y_{i,r,t} \qquad \forall i \in T \quad (10.26)$$

$$y_{i,r,t} \in \{0,1\} \qquad \forall i \in T, r, t \in A_i : r < t, t - r \leqslant \tau \quad (10.27)$$

Balancing Rest Times over Teams with IP

Ogryczak (2000) shows that equitable dominance between timetables D and D' can be identified by comparing their cumulative ordered aversion vectors. To this purpose, let $\vec{\Theta}_D = \big(\theta_1(D), \theta_2(D), \ldots, \theta_n(D)\big)$ be the cumulative ordered aversion vector where $\theta_i(D) = \sum_{j=1}^{i} f_j(D)$ for $i \in T$ (recall that $f_i(D) \leqslant f_{i+1}(D)$). Timetable D equitably dominates D' if, and only if, $\theta_i(D) \leqslant \theta_i(D')$ for all $i \in T$ with at least one strict inequality (Ogryczak, 2000). As an example, timetable D with $\vec{x}_D = (15, 10, 5)$ dominates D' with $\vec{x}_{D'} = (20, 10, 0)$ as we have $\vec{\Theta}_D = (15, 25, 30)$ and $\vec{\Theta}_{D'} = (20, 30, 30)$.

An important consequence is that all equitably-efficient timetables may be generated by enumerating all Pareto-efficient solutions with respect to $\min_{D \in \mathcal{D}} \vec{\Theta}_D$. To this purpose, Kostreva et al. (2004) propose constraints (10.28)-(10.30). Constraints (10.28) model auxiliary variables $d_{i,j}^+$ representing the upside deviation of the aversion of team i from the value of the unrestricted variable t_i. These auxiliary variables are then used in constraints (10.29) to model $\theta_i(D)$. For a correctness proof, we refer to Kostreva et al. (2004) and Ogryczak et al. (2008). Applying these constraints to our problem leads to the following IP-EQ model.

IP-EQ model

min $\vec{\Theta} = (\theta_1, \theta_2, \ldots, \theta_n)$

subject to

(10.20) − (10.27)

$$t_i + d_{i,j}^+ \geq f_i \qquad \forall i, j \in T \quad (10.28)$$

$$\theta_i = it_i + \sum_{j \in T} d_{i,j}^+ \qquad \forall i \in T \quad (10.29)$$

$$d_{i,j}^+ \geq 0 \qquad \forall i, j \in T \quad (10.30)$$

There are several approaches to generate Pareto-efficient solutions for the IP-EQ model. One approach is to use an aggregation function which maps $\vec{\Theta}_D$ into a single objective to be optimised. Kostreva et al. (2004) characterise aggregation functions that result in an equitably-efficient solution. One example of such an aggregation function is $\sum_{i \in T} w_i \theta_i(D)$ with $w_i = \big(n + (n - 2i + 1)\lambda\big)/n^2$ and λ a trade-off parameter in the range $]0, 1]$ (see Ogryczak (2000)). We will use this aggregation function in the computational results of Section 10.4.2 where we solve the IP-EQ model with Gurobi Optimiser 7.5.2 using

13 threads and 8 GB of RAM and a time limit of 3 hours.

Another approach is to treat the objectives hierarchically. Rawl's difference principle states that "social and economic inequalities are to be arranged so that they are to be of greatest benefit to the least-advantaged members of society" (Rawls and Kelly, 2001) and therefore suggests a *minimax* approach which minimises $\theta_1(D)$ first. The minimax approach, however, only concerns the aversion of the worst-off team (i.e., $\theta_1(D)$) and thereby misses remaining optimisation opportunities. Ogryczak (1997) proves that equitably-efficient solutions can be generated by the *lexicographic minimax* approach which optimises $\theta_i(D)$ in increasing order of i, without allowing to deteriorate previously optimised objectives. Utilitarian philosophers, on the other hand, would advocate the minisum approach which corresponds to minimisation of $\theta_n(D)$. An equitably-efficient solution can be generated by the *lexicographic minisum* approach which optimises $\theta_i(D)$ in decreasing order of i.

A Bi-Criteria Evolutionary Algorithm

The set of Pareto-efficient solutions for $\vec{\Theta}_D$ defines the entire set of equitably-efficient solutions. Unfortunately, direct optimisation of $\vec{\Theta}_D$ implies the use of an aggregation function or requires that the objectives are optimised hierarchically. As it is daunting, if not impossible, to define the preference of the league organiser prior with regard to the efficiency-equity trade-off, this section proposes a bi-criteria evolutionary algorithm generating a rich set of equitably-efficient compromise solutions. This approach is not only appealing from a computational point of view but also allows to visualise the trade-off in the efficiency-equity image space.

In particular, the bi-criteria evolutionary algorithm takes into account criterion efficiency via mean aversion $\mu_D = \sum_{i \in T} f_i(D)/n$ and distribution equity via the mean absolute difference MD_D:

$$\text{MD}_D = \frac{1}{2n^2} \sum_{i,j \in T} |f_i(D) - f_j(D)|. \tag{10.31}$$

Intuitively, MD_D expresses the expected difference in the aversion of two uniformly chosen teams.

Noteworthy, existing timetabling literature measures equity via Jain's fairness index $J_D = (\sum_{i=1}^n f_i(D))^2 / (n \sum_{i=1}^n f_i(D)^2)$ (see, e.g., Mühlenthaler and Wanka (2016) and Muklason et al. (2017)) or Gini's coefficient $G_D = \text{MD}_D / \mu_D$. Unfortunately, we cannot use Jain's index and Gini's coefficient as they do not comply with the concept of equitable-efficiency (see Ogryczak (2000)).

In order to minimise the ARTP, Van Bulck et al. (2019) propose

10.3. FOR ACADEMICS

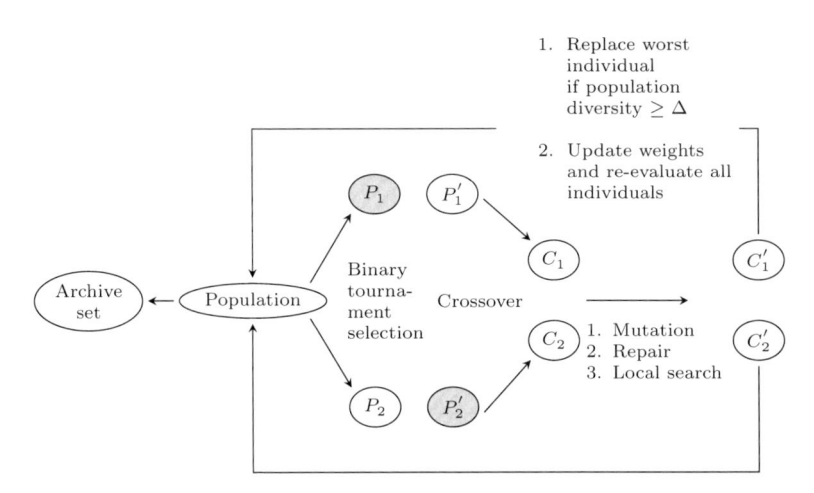

Figure 10.2: Illustration of the memetic algorithm with population management (MA|PM). Grey ellipses represent the parents selected by the binary tournament operator.

a tabu-search based algorithm. The key component in this algorithm consists of scheduling (or rescheduling) all home games of one team, which is modelled as a transportation problem. To fairly distribute rest times over teams, we propose a multi-objective genetic algorithm which includes a local search operator based on the algorithm of Van Bulck et al. (2019). Genetic algorithms which involve a local search operator are known as memetic algorithms; we refer to Jaszkiewicz et al. (2012) for an introduction on memetic algorithms. To cope with multiple objectives, our algorithm uses several ideas proposed by Ishibuchi and Murata (1998). The remainder of this section briefly summarises the main components of our algorithm. A general overview of the algorithmic flow is depicted in Figure 10.2.

Solution Representation The algorithm encodes a double round-robin timetable via a matrix where each cell (i, j) carries round $r_{i,j}$ on which home team $i \in T$ plays against away team $j \in T \setminus \{i\}$. We allow the algorithm not to plan a game by leaving the corresponding cell blank, but this results in a high cost P during the fitness evaluation.

Fitness of Individuals For each candidate timetable D, we count the total number of unscheduled games u_D and assign fitness $Pu_D + \mu_D + \lambda \text{MD}_D$. Ogryczak (2000) proves that any timetable which is optimal for this fitness function is equitably-efficient when $0 < \lambda \leqslant 1$.

At the beginning of each generation, the algorithm varies the search direction by uniformly choosing the trade-off parameter λ in the range $]0,1]$. This facilitates finding a diverse set of equitably-efficient solutions (see Ishibuchi and Murata (1998)).

Recombination and Mutation To improve solutions and to avoid getting trapped in local optima, our genetic algorithm varies candidate solutions via crossover and mutation. The crossover operator takes two parent solutions as input and generates a new offspring solution as follows. First, it draws a random number t between 1 and $n-1$. Next, it copies all home game assignments of the first t teams from the first parent, and all other home game assignments from the second parent. Since each game features one home team and one away team, this makes that all games are scheduled. We also consider a variant of this operator which copies away game assignments. After crossover, each offspring solution undergoes mutation. For each cell (i, j) the algorithm randomly decides independently whether cell value $r_{i,j}$ is to be mutated, in which case $r_{i,j}$ is replaced by a uniformly chosen value from $\{H_i \cap A_j\} \setminus \{r_{i,j}\}$.

Repair and Local Search As new offspring solutions may violate constraints *(C4)* and *(C5)*, we fully repair a timetable for all teams in a randomly chosen order by solving a transportation network which schedules or reschedules all home games of a chosen team. Van Bulck et al. (2019) explain how this transportation network can be adapted to optimise the ARTP. The fitness function used in our version of the algorithm, however, varies the value of λ to take into account distribution equity alongside criterion efficiency. Therefore, the fitness of a solution may deteriorate after solving the transportation network. We only accept the new solution if it improves the fitness value.

Population Management Genetic algorithms need a parent and survivor selection scheme to guide the evolution of the population. We create the initial population by repeatedly solving the IP model presented in Equations (10.20)-(10.24). At the beginning of each iteration, our algorithm uniformly chooses $0 < \lambda \leqslant 1$ and re-evaluates all individuals in the current population. Next, a binary tournament operator selects two parent solutions based on the re-evaluated fitness values. With a probability of p_c, the two parents' mate, in which case two off-springs are generated by crossover. In the other case the two off-springs are identical to the two parents. With a probability of p_m, the offspring solutions undergo mutation. After recombination and mutation, the local improvement heuristic is applied. Subsequently, for each offspring and each member of the population, the algorithm calculates a dissimilarity distance expressed in terms of the percentage of different game to round assignments. If the distance between the offspring and each member of the population is greater than the diversity parameter Δ, the offspring replaces the worst solution in the current population (for more details, see Sörensen and Sevaux (2006)).

Apart from the current population, our version of the memetic algorithm stores an archive set of equitably-efficient solutions. If an offspring solution is not equitably dominated by any other solution in the archive, the offspring solution is added to the archive from which we subsequently remove newly equitably dominated solutions. By using this archive set, no equitably-efficient solutions are lost (for more information, see Ishibuchi and Murata (1998)). Ideally, solutions in the archive set are close to the Pareto-efficient frontier and cover the entire mean-equity Pareto front.

Computational Setting The memetic algorithm is implemented in `C++`, compiled with `g++` 4.8.5, using optimisation flag -O3 and parallelised with OpenMP. The parameters of the algorithm were tuned with `irace` (López-Ibáñez et al., 2016), using the hypervolume indicator. To solve the transportation problems, we use an $\mathcal{O}(n^3)$ implementation of the Kuhn-Munkres algorithm. We run the genetic algorithm with a time-limit of 10 minutes, 12 cores, and 2 GB of RAM on a CentOS 7.4 GNU/Linux based system with an Intel E5-2680 processor running at 2.5 GHz.

10.4 For Practitioners

10.4.1 Trade-Offs Between Two Fairness Criteria

Every year, football fans and clubs eagerly await the announcement of the official timetable for the new season by the league organisers. This timetable is the result of a complex planning process and tries to consider wishes and requirements from the police, broadcasters, sponsors, etc. Since it is rarely revealed exactly what these requirements are, the media focus mostly on the fairness of the timetable.

In this section we discuss the number of breaks and the carry-over effect present in the official timetables of 10 main European football leagues: Belgium, England, France, Germany, Italy, the Netherlands, Portugal, Russia, Spain, and Ukraine (for other empirical studies on football timetables, we refer to Goossens and Spieksma (2012b) and Yi et al. (2020)). We studied all timetables from season 2009 − 2010 up to season 2018 − 2019. Table 10.7 lists the average carry-over effects value per team and average number of breaks per team for each season.

Looking at breaks, England has a strikingly high number, with the Netherlands coming second. It is safe to say that the timetabling process in these countries leaves a lot of room for improvement. In most other countries the number of breaks is low and quite stable; although Russia has had a few aberrant seasons, and France seems to have abandoned their timetables with two breaks per team since season 2014 − 2015.

Except for the last season, Spain displays the largest carry-over effects value among the 10 leagues. The main reason is that it uses a so-called canonical timetable. The canonical timetable has been (and to some extent still is) very popular in sports timetabling (Goossens and Spieksma, 2012b); however, it is also the worst possible timetable with respect to balancing carry-over effects (Lambrechts et al., 2017). The canonical timetable has also been used in Ukraine (all seasons) and Russia (7 of the 10 seasons).

Table 10.7 sketches a somewhat distorted picture since the number of teams in a league has an impact on the minimal number of breaks and the minimal carry-over effects value that can be attained. The leagues indeed do not feature the same number of teams: England, France, Italy, and Spain have 20 teams, the Netherlands and Germany have 18 teams, and Belgium and Russia have 16 teams. In Portugal, the number of teams increased from 16 to 18 in season 2014 − 2015, while in Ukraine, the number of teams dropped from 16 to 14 in that same season, and even to 12 in season 2016 − 2017. In order to allow a proper comparison, Figure 10.3 shows the average number of breaks per team on its horizontal axis. While this makes sense for breaks, it is much less the case for the average carry-over effect per team, since the minimal carry-over effects value increases non-linearly with the number of teams. Therefore, Figure 10.3 presents how close the carry-over effects value is to the maximal carry-over effects value for a league of that size (corresponding to an indexed value of 1) on its vertical axis. Since the maximal carry-over effects value is not known for double round robin tournaments, and since many timetables apply mirroring, we have focused on the first half of the season only.

The trade-off between balancing carry-over effects and minimising the total number of breaks is now apparent in Figure 10.3. The figure clearly shows the Netherlands and England completely neglect the number of breaks, and Spain, Russia, and Ukraine ignore the carry-over effects value. The Pareto-front, which corresponds to the best trade-off one can make, is also indicated in Figure 10.3. The Pareto-front is based on timetables from Italy, France, and Belgium. That means that for these seasons, no one does better with respect to breaks *and* carry-over. Although they make different choices with respect to the trade-off, these three countries are "getting the most value for their money". Portugal and Germany are not far away in most

Table 10.7: Overview of the carry-over effects value / breaks for 10 main European football leagues for seasons 2009 – 2010 till 2018 – 2019

Year	Belgium	England	France	Germany	Italy	Netherlands	Portugal	Russia	Spain	Ukraine
09/10	2383/42	2436/172	4548/40	3542/48	2974/66	1966/114	2307/42	8345/40	20657/54	9015/50
10/11	2441/42	2892/124	4730/40	5154/48	3306/64	1778/112	2231/42	8491/32	21044/54	9015/44
11/12	2571/42	2574/126	4740/40	4460/48	3144/66	1906/124	2287/42	9591/50	21044/54	9015/60
12/13	2663/42	2034/130	4680/48	4202/48	3124/62	1940/106	2307/42	8491/40	21044/54	9015/54
13/14	2265/66	2764/126	4536/40	3622/48	3508/64	1930/116	3962/42	1431/112	21044/54	9015/42
14/15	2423/44	2764/116	4536/48	4342/48	3508/62	1930/116	3962/48	1431/92	21044/54	5517/52
15/16	2201/44	2594/138	2258/84	4050/48	3232/64	1806/110	3842/48	8491/32	21044/54	5477/36
16/17	1545/52	2696/154	2510/74	4046/48	3124/64	2076/110	3854/48	4859/78	21044/54	3075/32
17/18	1477/48	2540/152	2416/66	3874/48	3160/72	2142/88	4058/48	8491/44	21044/58	3075/32
18/19	1513/48	2642/132	2498/70	3734/48	3268/64	2132/134	4330/48	8491/60	2598/72	3075/36

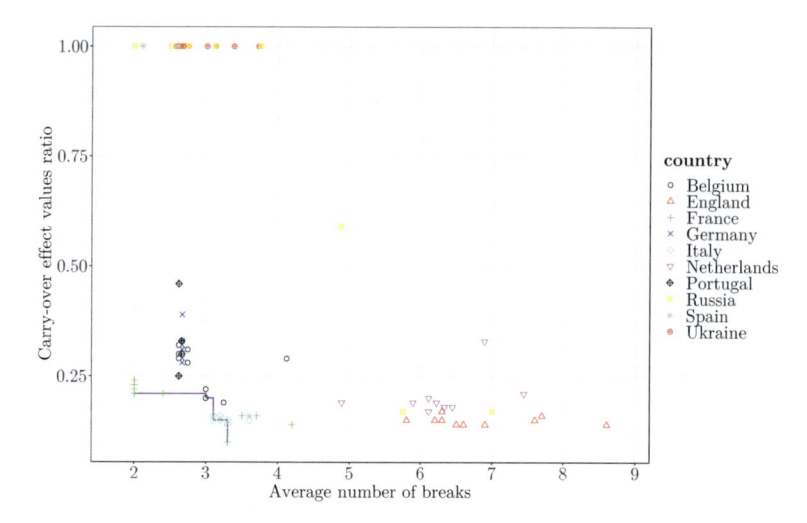

Figure 10.3: Trade-off between balancing carry-over effects and minimising breaks (each entry corresponds to one season in one league).

seasons, but leagues like England could get a much lower number of breaks for the same carry-over effects value. Likewise, Spain could drastically improve its balance of carry-over effects without incurring more breaks. It is interesting to see that while most leagues have their seasons in the same area of the plot, the league organisers in Russia make very different choices from one season to the next, i.e., sometimes opting for a very high carry-over effects value and small number of breaks, and sometimes vice versa.

10.4.2 Trade-Offs Between Efficiency and Equity

The "Liefhebbers Zaalvoetbal Cup" (LZV Cup) is a non-professional indoor football league founded in 2002. This league currently involves 548 teams, grouped divisions in 20 regions in Flanders (Belgium). In a division each team plays each other team once at home and once away in a double round-robin tournament. The league aims to attract teams that consist of friends, is open to all ages, and considers fair play of utmost importance. The games are played without referees because, according to the organisers, "referees are expensive, make mistakes,

10.4. FOR PRACTITIONERS

and invite players to explore the borders of sportsmanship"[2]. In this context it makes sense that their timetables also display a high level of fairness.

In this chapter we consider 3 divisions of the LZV Cup: two of these divisions have 15 teams, the other has 14 teams[3]. The season starts on September 1st and ends on May 31st, which corresponds to 273 rounds. A time-relaxed timetable is required in which no team has more than two games in a period of τ rounds or less. For the divisions with 15 teams, $\tau = 8$ whereas in the division with 14 teams $\tau = 9$. We use $p_r = (\tau - r)^2$ to denote the penalty incurred for every pair of consecutive games played by a team within a period of $r < \tau$ rounds. We refer to the total sum of penalties incurred by team i as the aggregated rest time penalty of i (ARTP_i); the sum of the ARTP_i values over all teams results in the ARTP score of the timetable.

In order to evaluate the fairness of a timetable, we use two decision criteria. First, we use the mean aggregated rest time penalty $\mu_D = ARTP/n$ to measure the criterion efficiency of timetable D. Second, we use the mean absolute difference $\text{MD}_D = \frac{1}{2n^2} \sum_{i=1}^{n} \sum_{j=1}^{n} |\text{ARTP}_i - \text{ARTP}_j|$ to measure the equity of timetable D with regard to the distribution of the rest time penalties over the teams. Intuitively, MD_D measures the expected difference in the aggregated rest time penalty of two randomly chosen teams.

The league organisers are confronted with the price of equity: by how much does the mean aggregated rest time penalty increase when we additionally consider the distribution of the penalties over the teams? After generating schedules manually for a number of years, the league organisers have moved to an integer programming formulation (based on constraints (10.20)-(10.24) described in Section 10.3.2). However, this approach only gives them one single solution per division and no insight on the trade-off between equity and efficiency. In this chapter we have developed an IP formulation IP-EQ and an evolutionary algorithm in order to produce alternative timetables and to shed light on the price of equity.

Figure 10.4 shows the Pareto-front with regard to criterion efficiency (μ_D) and distribution equity (MD_D) for each of the three divisions. Additionally, we draw a straight trade-off line through the solution with the best efficiency score and set the slope such that a 1% increase in equity yields a 1% increase in efficiency. Mühlenthaler

---

[2] see www.lzvcup.be [in Dutch]

[3] More details on these divisions can be found in Van Bulck et al. (2019), in which divisions 1 to 3 are referred to as benchmark instance #2, #3 and #4, respectively.

and Wanka (2016) argue that picking a timetable below this trade-off line might be an attractive option in real-world applications as the increase in fairness is larger than the decrease in efficiency.

When analysing Figure 10.4, we see that the official timetable is far above the Pareto frontier: more efficient and equitable timetables exist. The genetic algorithm is able to produce a rich set of equitably-efficient solutions which are well separated in the efficiency-equity (μ_D-MD_D) image space. This is remarkable as the genetic algorithm produces all its solutions in a single run with 10 minutes of computation time, whereas the IP-formulations generate only one solution after 3 hours of computation time.

Figure 10.4 also shows that the price of fairness can be quite different between divisions (compare, for example, a 1% reduction of MD_D in division 1 and 2). This motivates our approach to provide the league organiser with a rich set of equitably-efficient timetables. From this set, the league organiser may select a final timetable by, e.g., further analysing the rest times distribution.

As an example, Figure 10.5 shows the rest time penalty $ARTP_i$ for each of the 15 teams in division 2 for various timetables, where team 1 is the worst-off team in that timetable with respect to rest times, and team 15 has the most favourable rest times. The figure compares the timetable with the lowest maximal $ARTP_i$, the most equitable timetable, the most efficient timetable, and the timetable that was actually used in the LZV Cup competition. It would not be hard for the league organisers to convince the teams of the added value of our timetabling algorithms. Indeed, when comparing the most efficient timetable with the official one, we see that every team is better off. Unfortunately, the most efficient timetable does not consider the distribution of the rest times over the teams. The timetable with the lowest maximal $ARTP_i$ illustrates that the rest times of the worst-off team can be improved considerably, although this happens at the expense of nearly every other team when compared with the most efficient timetable. The most equitable timetable may work as a compromise solution as, except for team 2, mainly the teams with low $ARTP_i$ penalties pay the price to improve the situation for the worst-off team.

10.4. FOR PRACTITIONERS

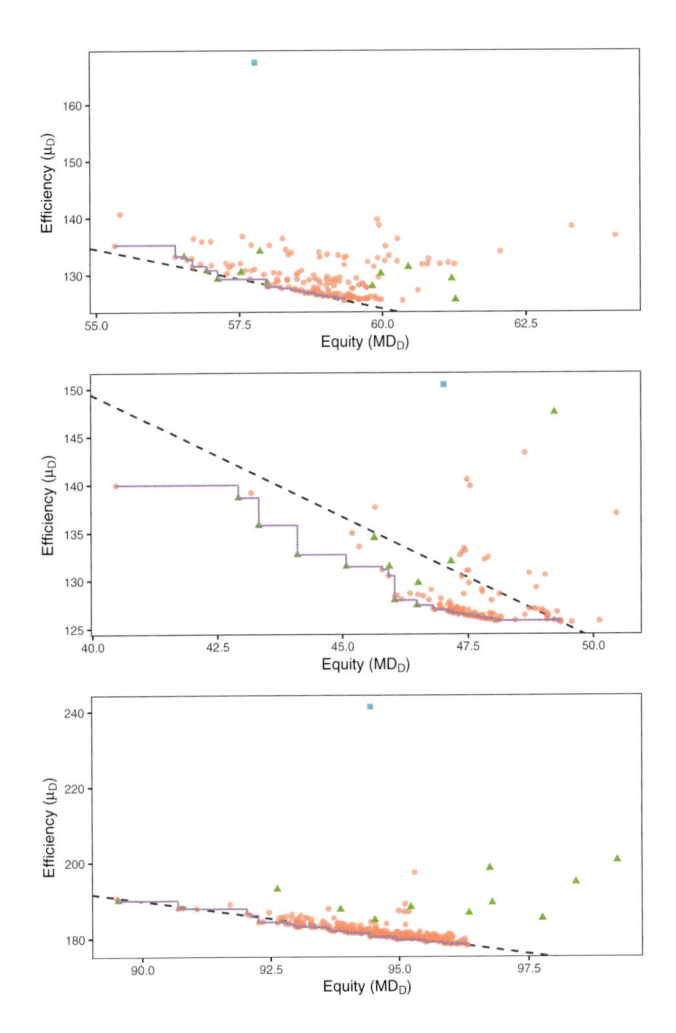

Figure 10.4: Pareto frontier for division 1 (top), 2 (middle), and 3 (bottom). Red circles represent the solutions found by the genetic algorithm, green triangles the solutions found with IP-EQ, and blue squares the official solutions. The Pareto frontier is indicated by the full line in purple, whereas the 1% trade-off line is indicated by the dashed line in grey.

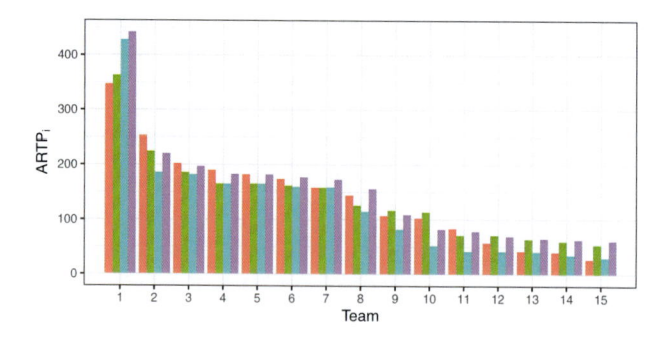

Figure 10.5: Aggregated rest time per team for the timetable with the lowest maximal ARTP ("MiniMax", red), most equitable timetable ("MiniMD", green), most efficient timetable ("MiniSum", blue), and the official timetable (purple). Teams are sorted in non-decreasing order of $ARTP_i$, which makes that the worst-off team (i.e., team 1) may be a different team depending on the timetable.

10.5 Conclusion

In round robin competitions it is often said that teams should not complain about the timetable, because "at the end of the day you have to play against everyone"[4]. This chapter has demonstrated that there is more to fairness in sports timetabling than that by discussing breaks, the carry-over effect, and rest times.

When confronted with multiple fairness issues, it may be difficult to get the best of both worlds. For instance, a better balance of the carry-over effects usually goes at the expense of a higher number of breaks. We developed a new method that generates timetables that offer a better trade-off than those currently available in the literature. We also looked at how the big European football competitions deal with this trade-off and found that France, Italy, and Belgium are making sensible choices; other leagues leave plenty of room for improvement.

Even a timetable with few unfair properties may be ill received when only a small number of teams carry most of this burden. In contrast to classic approaches which only minimise the total unfairness, we showed how to equitably distribute the unfair properties over the teams. We demonstrated the importance of balancing rest times between consecutive games in a real-life indoor football competition. We found that the distribution of rest times over the teams can be improved, although this results in a slightly shorter average rest time.

In conclusion, we want to stress the importance for league planners to work with a timetabling method that offers multiple and diverse timetables that are close to the Pareto front. Only in this way the league planners can make a thought-out choice on fairness trade-offs in sports timetabling.

Acknowledgement

The authors wish to thank Prof. Mario Guajardo for discussing fairness issues in sports, as well as the participants of the Fairness in Sports workshops (April 12th, 2018, Ghent, Belgium and June 5th, 2018, London, UK) who indirectly contributed to this chapter.

---

[4]This was for instance said by Juan Mata (Manchester United), after learning that his team had to open the 2019-2020 season with a challenging game against Chelsea, see www.manutd.com/en/news/detail/juan-mata-previews-man-utd-v-chelsea-to-start-new-premier-league-season.

Bibliography

Anderson, I. (1999). Balancing carry-over effects in tournaments. In Holroyd, F., Quinn, K., Rowley, C., and Webb, B., editors, *Combinatorial Designs and Their Applications*, pages 1–16, Boca Raton. CRC Research Notes in Mathematics. Chapman & Hall.

Bengtsson, H., Ekstrand, J., and Hägglund, M. (2013). Muscle injury rates in professional football increase with fixture congestion: an 11-year follow-up of the UEFA Champions League injury study. *British Journal of Sports Medicine*, 47:743–747.

Briskorn, D. and Knust, S. (2010). Constructing fair sports league schedules with regard to strength groups. *Discrete Applied Mathematics*, 158:123–135.

Çavdaroğlu, B. and Atan, T. (2019). Integrated break and carryover minimization problem in round robin tournament. In *MathSport International 2019 Conference Proceedings*, pages 25–32.

De Werra, D. (1980). Geography, games, and graphs. *Discrete Applied Mathematics*, 2:327–337.

De Werra, D. (1981). Scheduling in sports. *Studies on Graphs and Discrete Programming*, 11:381–395.

Forrest, D. and Simmons, R. (2005). New issues in attendance demand: The case of the English football league. *Journal of Sports Economics*, 7(3):247–266.

Friedman, M. (1975). *There's No Such Thing As a Free Lunch*. Open Court Publishing Company (US).

Goossens, D. R. and Spieksma, F. C. R. (2012a). The carryover effect does not influence football results. *Journal of Sports Economics*, 13(3):288–305.

Goossens, D. R. and Spieksma, F. C. R. (2012b). Soccer schedules in Europe: an overview. *Journal of Scheduling*, 15(5):641–651.

Guedes, A. C. and Ribeiro, C. C. (2011). A heuristic for minimizing weighted carry-over effects in round robin tournaments. *Journal of Scheduling*, 14(6):655–667.

Günneç, D. and Demir, E. (2019). Fair-fixture: Minimizing carry-over effects in football leagues. *Journal of Industrial & Management Optimization*, 15(4):1565–1577.

Henz, M., Müller, T., and Thiel, S. (2004). Global constraints for round robin tournament scheduling. *European Journal of Operational Research*, 153(1):92–101.

Ishibuchi, H. and Murata, T. (1998). A multi-objective genetic local search algorithm and its application to flowshop scheduling. *IEEE Transactions on Systems, Man, and Cybernetics, Part C (Applications and Reviews)*, 28:392–403.

Januario, T., Urrutia, S., Celso, C. R., and de Werra, D. (2016). Edge coloring: A natural model for sports scheduling. *European Journal of Operational Research*, 254:1–8.

Jaszkiewicz, A., Ishibuchi, H., and Zhang, Q. (2012). Multiobjective memetic algorithms. In *Handbook of Memetic Algorithms*, pages 201–217. Springer.

Karsu, O. and Morton, A. (2015). Inequity averse optimization in operational research. *European Journal of Operational Research*, 245:343–359.

Kidd, M. (2010). A tabu-search for minimising the carry-over effects value of a round-robin tournament. *ORiON*, 26(2):125–141.

Kostreva, M. M., Ogryczak, W., and Wierzbicki, A. (2004). Equitable aggregations and multiple criteria analysis. *European Journal of Operational Research*, 158:362–377.

Lambrechts, E., Ficker, A. M. C., Goossens, D. R., and Spieksma, F. C. R. (2017). Round-robin tournaments generated by the circle method have maximum carry-over. *Mathematical Programming*, 172:270–302.

López-Ibáñez, M., Dubois-Lacoste, J., Cáceres, L. P., Birattari, M., and Stützle, T. (2016). The irace package: Iterated racing for automatic algorithm configuration. *Operations Research Perspectives*, 3:43–58.

Miyashiro, R. and Matsui, T. (2006). Minimizing the carry-over effects value in a round-robin tournament. In *Proceedings of the 6th International Conference on the Practice and Theory of Automated Timetabling*, pages 460–463.

Mühlenthaler, M. and Wanka, R. (2016). Fairness in academic course timetabling. *Annals of Operations Research*, 239:171–188.

Muklason, A., Parkes, A. J., Özcan, E., McCollum, B., and McMullan, P. (2017). Fairness in examination timetabling: Student preferences and extended formulations. *Applied Soft Computing*, 55:302–318.

Ogryczak, W. (1997). On the lexicographic minimax approach to location problems. *European Journal of Operational Research*, 100:566–585.

Ogryczak, W. (2000). Inequality measures and equitable approaches to location problems. *European Journal of Operational Research*, 122:374–391.

Ogryczak, W., Wierzbicki, A., and Milewski, M. (2008). A multicriteria approach to fair and efficient bandwidth allocation. *Omega*, 36:451–463.

Rawls, J. and Kelly, E. (2001). *Justice as fairness: A restatement*. Harvard University Press, Cambridge (US).

Régin, J.-C. (2001). Minimization of the number of breaks in sports scheduling problems using constraint programming. *DIMACS Series in Discrete Mathematics and Theoretical Computer Science*, 57:115–130.

Russell, K. (1980). Balancing carry-over effects in round robin tournaments. *Biometrika*, 67(1):127–131.

Sörensen, K. and Sevaux, M. (2006). MA|PM: memetic algorithms with population management. *Computers & Operations Research*, 33:1214–1225.

Suksompong, W. (2016). Scheduling asynchronous round-robin tournaments. *Operations Research Letters*, 44:96–100.

Trick, M. A. (2000). A schedule-then-break approach to sports timetabling. In *Proceedings of the 3rd International Conference on the Practice and Theory of Automated Timetabling*, pages 242–253.

Van Bulck, D., Goossens, D., Schönberger, J., and Guajardo, M. (2020). RobinX: a three-field classification and unified data format for round-robin sports timetabling. *European Journal of Operational Research*, 270:568–580.

Van Bulck, D., Goossens, D. R., and Spieksma, F. C. R. (2019). Scheduling a non-professional indoor football league: a tabu search based approach. *Annals of Operations Research*, 275:715–730.

Yi, X., Goossens, D., and Nobibon, F. T. (2020). Proactive and reactive strategies for football league timetabling. *European Journal of Operational Research*, 282(2):772–785.